Principles of Astrophysics

Principles of Astrophysics

Mervin Williamson

Willford Press
www.willfordpress.com

Published by Willford Press,
118-35 Queens Blvd., Suite 400,
Forest Hills, NY 11375, USA

ISBN: 978-1-68285-734-2

Cataloging-in-Publication Data

Principles of astrophysics / Mervin Williamson.
p. cm.
Includes bibliographical references and index.
ISBN 978-1-68285-734-2
1. Astrophysics. 2. Physics. I. Williamson, Mervin.
QB461 .P75 2019
523.01--dc23

For information on all Willford Press publications
visit our website at www.willfordpress.com

Willford Press

Printed in China.

Contents

Preface VII

Chapter 1 **Astrophysics: An Introduction** **1**

i. Solar Physics 4
ii. Observational Astrophysics 4
iii. Theoretical Astrophysics 6
iv. Nuclear Astrophysics 11

Chapter 2 **Orbital Mechanics** **24**

i. Kepler's Laws of Planetary Motion 39
ii. Orbital Velocity 51
iii. Patched Conic Approximation 54
iv. Gravity Assist 58
v. Hohmann Transfer Orbit 63
vi. Geostationary Transfer Orbit 70

Chapter 3 **Radiation and Matter Interactions** **74**

i. Microwave Interactions 74
ii. Infrared Interactions 74
iii. Visible Light Interactions 75
iv. Ultraviolet Interactions 75
v. X-rays Interactions 76
vi. Photon Interactions 76
vii. Electron Interactions 79
viii. Positron Interactions 81

Chapter 4 **The Sun** **83**

i. Solar Core 86
ii. Corona 89
iii. Photosphere 90
iv. Chromosphere 92
v. Solar Cycle 93
vi. Solar Dynamo 96

Chapter 5 **The Solar System** **106**
i. Formation and Evolution of the Solar System 125
ii. Ring System 144
iii. Planetary System 151
iv. Small Solar System Body 164
v. Nice Model 167

Chapter 6 **Planet and its Types** **170**
i. Terrestrial Planet 185
ii. Giant Planet 187
iii. Exoplanet 195
iv. Dwarf Planet 204

Chapter 7 **Understanding Galaxies** **210**
i. Galaxy Formation and Evolution 222
ii. Galaxy Morphological Classification 224
iii. Milky Way 228
iv. Galactic Centre 231
v. Interacting Galaxy 232
vi. Magellanic Clouds 235

Permissions

Index

Preface

Astrophysics is a branch of astronomy that uses the principles of physics and chemistry to determine the nature of astronomical objects. Stars, galaxies, the cosmic microwave background, interstellar medium, etc. are studied in astrophysics. The emissions from such astronomical objects are analyzed and their properties of luminosity, temperature, chemical composition and density are also examined. The study of galaxy formation and evolution, dark matter, dark energy, stellar evolution and dynamics, etc. are also within the scope of this field. The approaches to the study of astronomical objects and astronomical phenomena fall under the two categories of observational and theoretical astrophysics. Observations in astrophysics are made by studying the electromagnetic radiation using radio, optical, space-based telescopes, etc. Neutrino and gravitational wave observatories are modern detectors in observational astrophysics. Theoretical astrophysics uses analytical models and numerical simulations to develop a comprehension of space phenomena. This textbook provides comprehensive insights into the field of astrophysics. It elucidates the concepts and innovative models around prospective developments with respect to this area of study. In this book, constant effort has been made to make the understanding of the difficult concepts as easy and informative as possible, for the readers.

To facilitate a deeper understanding of the contents of this book a short introduction of every chapter is written below:

Chapter 1- Astrophysics is a branch of astronomy. It is concerned with the study of the nature of astronomical objects. The objects under study in this field are the Sun, stars, planets, galaxies, and interstellar medium, cosmic microwave background, among others. This is an introductory chapter, which will discuss briefly all the significant concepts of solar physics, observational astrophysics, theoretical astrophysics and nuclear astrophysics.

Chapter 2- The field of study concerned with the application of celestial mechanics and ballistics to the motion of rockets and spacecraft is known as orbital mechanics. Some of the topics central to the development of orbital mechanics such as orbital velocity, Kepler's laws of planetary motion, gravity assist, Hohmann transfer orbit, geostationary transfer orbit, etc. have been covered in extensive detail in this chapter.

Chapter 3- An understanding of the interactions between matter and radiation can be developed from the study of microwave, infrared, visible, ultraviolet and X-ray interactions with matter as well as from the analysis of how photons, electrons and positrons interact with matter. This chapter has been carefully written to provide a comprehensive understanding of all aspects of interactions between matter and radiation.

Chapter 4- The sun in the center of the solar system. It is a sphere of hot plasma. The aim of this chapter is to provide an overview of the Sun. It includes detailed analysis of the solar core, corona, photosphere, chromosphere, solar cycle and solar dynamo.

Chapter 5- The Solar System is a gravitationally bound system comprising of the Sun and the planets. Astrophysics studies the solar system, with an emphasis on its formation and evolution, ring system, planetary system, etc. which have been discussed in this chapter.

Chapter 6- A planet is a celestial body orbiting a star. Planets can be of different types, such as terrestrial planets, gas giants, ice giants, exoplanets and dwarf planets, on the basis of size, composition or distance from the Sun. This chapter explores these varied types of planets and their related attributes.

Chapter 7- A galaxy is a system of stars and stellar remnants, dust, interstellar gas and dark matter. Some of the issues addressed in astrophysics regarding the study of galaxies are galaxy formation and evolution, interacting galaxy, galaxy classification, magellanic clouds, etc. which have been covered in this chapter.

I owe the completion of this book to the never-ending support of my family, who supported me throughout the project.

Mervin Williamson

1

Astrophysics: An Introduction

Astrophysics is a branch of astronomy. It is concerned with the study of the nature of astronomical objects. The objects under study in this field are the Sun, stars, planets, galaxies, and interstellar medium, cosmic microwave background, among others. This is an introductory chapter, which will discuss briefly all the significant concepts of solar physics, observational astrophysics, theoretical astrophysics and nuclear astrophysics.

Astrophysics is the study of the physical makeup and behaviour of celestial objects. Astrophysics and astronomy are not sharply distinguished. If one wishes to emphasize the application of physical laws or the physical understanding of astronomical objects, then the term astrophysics is used. The discipline of astrophysics involves the observation and analysis of planets and stars (including our sun), binary stars, clusters of stars, interstellar material, galaxies, quasars or anything else that is part of the universe. From a combination of theory and observation, astrophysicists deduce physical characteristics such as chemical composition, age, rotation rate, magnetic fields, space motions and so on. This information is then applied toward understanding how celestial objects came into being, how they change with time, how they will end up, and what forces and physical processes are involved.

Astrophysicists use many branches of physics: nuclear physics to study power-generation in stars; atomic physics to understand the spectra of stars and gaseous nebulae; and gas laws and magnetic theory to probe starspots and flares on star surfaces. Gravitational physics is fundamental to understanding planetary and binary-star orbits, the rotation of our galaxy and the expansion of the universe. Other sciences also make important contributions; for example, chemistry is needed to study atomic and molecular reactions in interstellar space; geology, geophysics and meteorology contribute to studies of planets. Technical skills in optics, electronics and computers are needed to collect astrophysical data and compare it to theory. Several fields of engineering contribute to the construction of astronomical telescopes (like the Canada-France-Hawaii Telescope), light detectors and spacecraft.

Radiation

Information about astronomical objects arrives in the form of electromagnetic radiation, i.e., light waves or photons. The most energetic form is gamma radiation, followed by X-rays, ultraviolet radiation, visible light (violet through red), infrared light, millimetre waves and radio waves, which have the least energy. Optical, X-ray, radio and other telescopes are required to measure the full range of radiation; only optical and radio radiation can be measured from the surface of the earth, because the atmosphere is opaque to the others. Photometry, polarimetry and spectroscopy are techniques for analysing light.

Photometry

Photometry measures brightness. An object's brightness as we see it depends on how far away the object is, how bright the object actually is and how much obscuring material lies along the line of sight. Hotter stars emit more blue than red light, compared to cooler stars. Starlight collected with a telescope is measured by a light detector; the relative amount of light in different colours is established by placing colour filters in front of the detector. The DDO (David Dunlap Observatory) colour system, designed by R.D. McClure and S. Van den Bergh, has been widely used. X-ray photometry can be used to study the outermost parts of a star, the chromospheres and coronae. Stellar eclipses and pulsation are also studied by photometry.

Polarimetry

Polarimetry measures the orientation of light waves. In natural-source radiation, waves are seen in all possible orientations and the light is unpolarized. If a magnetic field exists in the material emitting the light, the orientation becomes ordered and the light is polarized. Polarization also occurs when light is reflected or scattered, as starlight is by interstellar material. The polarization of visible light is detected by placing a Polaroid or Nichol prism in the light beam. In many cases, astronomical sources show only partial polarization (under 10%) if any. Magnetic fields are known to be strong in some types of galaxies because of the strongly polarized light and radio emissions.

Spectroscopy

Spectroscopy measures the detailed features in the electromagnetic spectrum. Most of these features are very narrow colour bands called spectral lines. They normally have less light than adjacent spectral regions, and they arise from the energy levels of the atoms in the celestial object. In the spectrum of a typical star, for instance, some of the more commonly seen spectral lines are produced by the elements hydrogen, sodium, magnesium, calcium, titanium, iron and nickel. Spectroscopic analysis of the strengths and shapes of spectral lines gives us information about the chemical composition, temperature, pressure, magnetic fields, convective motions, rotation rates and velocities of approach or recession of the object under study. Canadian researchers have done pioneering work in studies of stars, galaxies and quasars using spectroscopy.

- Goals of Astrophysics

Astrophysicists seek to understand the universe and our place in it. At NASA, the goals of astrophysics are "to discover how the universe works, explore how it began and evolved, and search for life on planets around other stars," according NASA's website.

NASA states that those goals produce three broad questions:

- How does the universe work?

- How did we get here?
- Are we alone?

It Began with Newton

While astronomy is one of the oldest sciences, theoretical astrophysics began with Isaac Newton. Prior to Newton, astronomers described the motions of heavenly bodies using complex mathematical models without a physical basis. Newton showed that a single theory simultaneously explains the orbits of moons and planets in space and the trajectory of a cannonball on Earth. This added to the body of evidence for the (then) startling conclusion that the heavens and Earth are subject to the same physical laws.

Perhaps what most completely separated Newton's model from previous ones is that it is predictive as well as descriptive. Based on aberrations in the orbit of Uranus, astronomers predicted the position of a new planet, which was then observed and named Neptune. Being predictive as well as descriptive is the sign of a mature science, and astrophysics is in this category.

Milestones in Astrophysics

Because the only way we interact with distant objects is by observing the radiation they emit, much of astrophysics has to do with deducing theories that explain the mechanisms that produce this radiation, and provide ideas for how to extract the most information from it. The first ideas about the nature of stars emerged in the mid-19th century from the blossoming science of spectral analysis, which means observing the specific frequencies of light that particular substances absorb and emit when heated. Spectral analysis remains essential to the triumvirate of space sciences, both guiding and testing new theories.

Early spectroscopy provided the first evidence that stars contain substances also present on Earth. Spectroscopy revealed that some nebulae are purely gaseous, while some contain stars. This later helped cement the idea that some nebulae were not nebulae at all they were other galaxies!

In the early 1920's, Cecilia Payne discovered, using spectroscopy, that stars are predominantly hydrogen (at least until their old age). The spectra of stars also allowed astrophysicists to determine the speed at which they move toward or away from Earth. Just like the sound a vehicle emits is different moving toward us or away from us, because of the Doppler shift, the spectra of stars will change in the same way. In the 1930's, by combining the Doppler shift and Einstein's theory of general relativity, Edwin Hubble provided solid evidence that the universe is expanding. This is also predicted by Einstein's theory, and together form the basis of the Big Bang Theory.

Also in the mid-19th century, the physicists Lord Kelvin (William Thomson) and Gustav Von Helmholtz speculated that gravitational collapse could power the sun, but eventually realized that energy produced this way would only last 100,000 years. Fifty years later, Einstein's famous $E=mc^2$ equation gave astrophysicists the first clue to what the true source of energy might be (although it turns out that gravitational collapse does play an important role). As nuclear physics, quantum mechanics and particle physics grew in the first half of the 20th century, it became possible to formulate theories for how nuclear fusion could power stars. These theories describe how stars form, live and die, and successfully explain the observed distribution of types of stars, their spectra, luminosities, ages and other features.

Astrophysics is the physics of stars and other distant bodies in the universe, but it also hits close to home. According to the Big Bang Theory, the first stars were almost entirely hydrogen. The nuclear fusion process that energizes them smashes together hydrogen atoms to form the heavier element helium. In 1957, the husband-and-wife astronomer team of Geoffrey and Margaret Burbidge, along with physicists William Alfred Fowler and Fred Hoyle, showed how, as stars age, they produce heavier and heavier elements, which they pass on to later generations of stars in ever-greater quantities. It is only in the final stages of the lives of more recent stars that the elements making up the Earth, such as iron (32.1 percent), oxygen (30.1 percent), silicon (15.1 percent), are produced. Another of these elements is carbon, which together with oxygen, make up the bulk of the mass of all living things, including us. Thus, astrophysics tells us that, while we are not all stars, we are all stardust.

Solar Physics

A branch of physics which focus on study of sun is called solar physics. Sun is considered as a special star since it is the one which provides most of the energy to earth. It is also the only star which can resolve the spatial scale.

Many of the properties of sun is still a mystery and study of it is thus of great importance. A close track on the sun is really important since atmosphere and activities of sun have a great deal of impact on earth.

The study of various properties of sun ranges back from ancient times. The ancient calendars were based on solar eclipse and sunspots. Ancient people also used solar observations for timekeeping and for navigation purposes. It was based on solar observations that Nicolas Copernicus put forward heliocentric model. In modern day research on sun is to study the sunspot, coronal heat problem etc.

Solar physics is related to many other branches of physics like atomic and molecular physics, astrophysics etc. The sun acts as a laboratory for the study of plasma.

Observational Astrophysics

The Pleiades, an open cluster of stars observed in the constellation of Taurus

Astrophysical observations are generally made in various segments of the electromagnetic spectrum, as indicated below.

- Optical astronomy is the oldest kind of astronomy, involving the observation of celestial objects in the visible range of light. The instrument most commonly used is the telescope, paired with a charge-coupled device or spectroscope. The Earth's atmosphere interferes somewhat with optical observations, so image quality is improved by the use of adaptive optics and space telescopes. By studying the spectra of objects in the night sky, researchers are able to determine the chemical composition of stars, galaxies, and nebulae.
- In radio astronomy, scientists study radiation with a wavelength greater than a few millimeters. The instruments used are very large radio telescopes. Radio waves are usually emitted by cold objects, including interstellar gas and dust clouds. The cosmic microwave background radiation is thought to be the redshifted light from the Big Bang. Pulsars were first detected at microwave frequencies.
- Infrared astronomy involves the study of radiation of wavelengths longer than the wavelengths of visible light but shorter than those of radio waves. Infrared observations are usually made with telescopes similar to the usual optical telescopes. Objects colder than stars (such as planets) are normally studied at infrared frequencies.
- Ultraviolet, X-ray, and gamma ray astronomy involve the study of very energetic processes, such as binary pulsars, black holes, magnetars, and many others. These types of radiation are blocked by the Earth's atmosphere to a large extent. To observe celestial objects and phenomena in these regions of the electromagnetic spectrum, scientists rely on space-based telescopes and ground-based imaging air Cherenkov telescopes (IACT). Observatories of the first type include RXTE, the Chandra X-ray Observatory, and the Compton Gamma Ray Observatory. Examples of IACTs are the High Energy Stereoscopic System (H.E.S.S.) and the MAGIC telescope.

Besides studying electromagnetic radiation from distant objects, astrophysicists also look for such things as gravitational waves, neutrinos, and cosmic rays. A few gravitational wave observatories have been constructed, but gravitational waves are extremely difficult to detect. Neutrino observatories have also been built, primarily to study the Sun. Cosmic rays, consisting of very high energy particles, can be observed striking the Earth's atmosphere.

Observations can also vary by timescale. Most optical observations take minutes to hours, so phenomena that occur faster than that cannot be readily observed. Also, historical data on some objects is available spanning centuries or millennia. On the other hand, radio observations may look at events on a millisecond timescale (millisecond pulsars) or combine years of data (pulsar deceleration studies). The information obtained from these different timescales is very different.

Study of the Sun occupies a special place in observational astrophysics. Given the tremendous distances of all other stars, the Sun can be observed at a level of detail unparalleled by any other star. Human understanding of the Sun serves as a guide to the understanding of other stars.

The topic of how stars change, or stellar evolution, is often modeled by placing the varieties of star types in their respective positions on what is called the Hertzsprung-Russell diagram. This diagram can be viewed as representing the state of a stellar object from birth to destruction.

The material composition of astronomical objects can often be examined using:

- Spectroscopy
- Radio astronomy
- Neutrino astronomy (future prospects)

Theoretical Astrophysics

Theoretical astrophysicists endeavor to create theoretical models and figure out the observational consequences of those models. This approach helps observers look for data that can confirm or refute a model, or helps them choose between several alternate models.

Theorists also try to generate or modify models to take into account new data. In the case of an inconsistency, the general tendency is to make minimal modifications to a model to fit the data. In some cases, a large amount of inconsistent data over time may lead to total abandonment of a model.

Theoretical astrophysicists use a variety of tools, including analytical models (such as polytropes to approximate the behavior of a star) and computational numerical simulations. Each offers some advantages. Analytical models of a process are generally better at giving insights into the heart of what is going on. Numerical models can reveal the existence of phenomena and effects that cannot otherwise be seen.

Topics studied by theoretical astrophysicists include: Stellar dynamics and evolution; galaxy formation; large-scale structure of matter in the Universe; origin of cosmic rays; general relativity and physical cosmology, including string cosmology and astroparticle physics. Astrophysical relativity serves as a tool to gauge the properties of large-scale structures in which gravitation plays a significant role in physical phenomena, and as a basis for understanding black holes and gravitational waves.

Some widely accepted and studied theories and models in astrophysics, now included in the Lambda-CDM model are the Big Bang, cosmic inflation, dark matter, and fundamental theories of physics.

A few examples of this process:

Physical process	Experimental tool	Theoretical model	Explains/predicts
Gravitation	Radio telescopes	Self-gravitating system	Emergence of a star system
Nuclear fusion	Spectroscopy	Stellar evolution	How the stars shine and how metals formed
The Big Bang	Hubble Space Telescope, COBE	Expanding universe	Age of the Universe
Quantum fluctuations		Cosmic inflation	Flatness problem
Gravitational collapse	X-ray astronomy	General relativity	Black holes at the centre of Andromeda galaxy
CNO cycle in stars			

Dark matter and dark energy are the current leading topics in astrophysics, as their discovery and controversy originated during the study of galaxies.

Computational Astrophysics

Computational astrophysics is the use of numerical methods to solve research problems in astrophysics on a computer.

Numerical methods are used whenever the mathematical model describing an astrophysical system is too complex to solve analytically (with pencil and paper). Today, it is difficult to find examples of research problems that do *not* use computation.

Computational Versus Analytic Methods

Solutions generated by numerical methods are generally only approximations to the exact solution of the underlying equations. However, much more complex systems of equations can be solved numerically than can be solved analytically. Thus, approximate solutions to the exact equations found by numerical methods often provide far more insight than exact solutions to approximate equations that can be solved analytically. For example, time-dependent numerical solutions of fluid flow in three dimensions can exhibit behavior which is not expected from one-dimensional analytic solutions to the steady-state (time independent) equations.

The increase in computing power in the last few decades has meant that an increasingly larger share of problems in astrophysics can be solved on a desktop computer. However, the most computationally intensive problems in astrophysics are still limited by the memory and floating-point speed of the largest high-performance computer (HPC) systems available (e.g. computers on the top500 list). The use of HPC is necessary to maximize the spatial, temporal, or frequency resolution of solutions, to include more physics, or when large parameter surveys are required to understand the statistics.

Examples

Traditionally, the most important applications of computation in astrophysics have been in the areas of:

- Stellar structure and evolution.

The internal structure and evolution of stars with different masses and chemical composition was largely mapped out in the 1960's using numerical methods to solve the equations of stellar structure. Today, the frontiers of research include calculating multidimensional stellar models of rapidly rotating stars, modeling the effects of hydrodynamical processes such as convection from first principles, and understanding how stars generate magnetic fields through dynamo processes.

- Radiation transfer and stellar atmospheres.

Without oversimplifying assumptions, computational methods are required to calculate the propagation of light through the outer layers of a star, including its interaction with matter through absorption, emission, and scattering of photons. The calculation of cross sections for the interaction of light with matter for astrophysically relevant ions is itself a challenging computational problem.

The construction of such stellar atmosphere models share many challenges with radiation transfer problems in other systems such as planets, accretion disks, and the interstellar medium. Modern calculations improve the frequency resolution, include a better treatment of opacities, and can treat non-hydrostatic atmospheres.

- Astrophysical fluid dynamics.

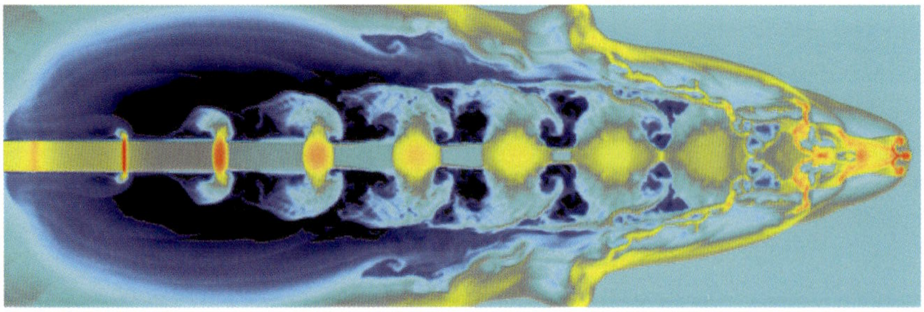

Figure: Density in a 2D simulations of the propagation of a pulsed, supersonic, protostellar jet

The dynamics of most of the visible matter in the universe can be treated as a compressible fluid. Time-dependent and multidimensional solutions to the fluid equations, including the effects of gravitational, magnetic, and radiation fields, require numerical methods. A vast range of problems are addressed in this way, from convection and dynamo action in stellar and planetary interiors, to the formation of galaxies and the large scale structure of the universe.

- Planetary, stellar, and galactic dynamics.

It is well known that Newton's Laws of Motion for some number of particles N interacting through their mutual gravitational attraction do not in general have an analytic solution for $N > 2$. Thus, to compute the orbits of planets in the solar system, or of stars in the Galaxy, numerical methods are required. The most challenging problems today include accurate integration of the orbits of the planets over the age of the solar system, studying the dynamics of globular clusters including the effect of stellar evolution and the formation of binaries, studying galaxy mergers and interaction, and computing structure formation in the universe through the gravitational clustering of collisionless dark matter.

Numerical Methods

The diverse set of mathematical models encountered in astrophysics means that a very wide range of numerical methods are necessary. These range from basic methods for linear algebra, nonlinear root finding, and ordinary differential equations (ODEs), to more complex methods for coupled partial differential equations (PDEs) in multi-dimensions, topics which span the entire content of reference books such as Numerical Recipes.

However, there are several numerical methods used in astrophysics that deserve special mention, either because they have such wide use in astrophysics, or because astrophysicists have made significant contributions to their development. These methods are considered in the following subsections.

Stellar Structure Codes

The equations of stellar structure are a set of ODEs which define a classic two-point boundary value problem. Analytic solutions to an approximate system of equations exist in special cases (*polytropes*) but these are of limited application to real stars. Numerical solutions to the full system of equations were first computed using *shooting methods*, in which boundary conditions are guessed at the centre and surface of the star, the equations are integrated outwards and inwards, and matching conditions are used at some interior point to select the full solution. Shooting methods are laborious and cumbersome; today modern stellar structure codes use relaxation schemes which find the solution to the finite-difference form of the stellar structure equations by finding the roots of coupled, non-linear equations at each mesh point. A good example of a public code that uses a relaxation scheme is the EZ-code, based on Eggleton's variable mesh method. The evolution of stars are then computed by computing stellar models at discrete time intervals, with the chemical composition of the star modified by nuclear reactions in the interior.

Radiative Transfer Codes

Calculating the emergent intensity from an astrophysical system requires solving a multidimensional integral-differential equation, along with level population equations to account for the interaction of the radiation with matter. In general the solution is a function of two angles, frequency, and time. Even in static, plane-parallel atmospheres, the problem is two-dimensional (one angle and frequency). However, the most challenging aspect of the problem is that scattering couples the solutions at different angles and frequencies. As in the stellar structure problem, relaxation schemes are used to solve the finite difference form of the transfer equations, although specialized iteration techniques are necessary to accelerate convergence. Monte-Carlo methods, which adopt statistical techniques to approximate the solution by following the propagation of many photon packets, are becoming increasingly important. The problem of line-transfer in a moving atmosphere (stellar wind) is especially challenging, due to non-local coupling introduced by Doppler shifts in the spectrum.

N-body Codes

There are two tasks in an N-body code: integrating the equations of motion (*pushing particles*), and computing the gravitational acceleration of each particle. The former requires methods for integrating ODEs. Modern codes are based on a combination of high-order difference approximations (e.g. Hermite integrators), and symplectic methods (which have the important property of generating solutions that obey Liouville's Theorem, i.e. that preserve the volume of the solution in phase space). Symplectic methods are especially important for long term time integration, because they control the accumulation of truncation error in the solution.

Calculation of the gravitational acceleration is challenging because the computational cost scales as $N(N-1)$, where N is the number of particles. For small N, direct summation can be used. For moderate N (currently N~105–6), special purpose hardware (e.g. GRAPE boards) can be used to accelerate the evaluation of 1/r2 necessary to compute the acceleration by direct summation. Finally, for large N (currently N≥109–10), tree-methods are used to approximate the force from distant particles.

Codes for Astrophysical Fluid Dynamics

Solving the equations of compressible gas dynamics is a classic problem in numerical analysis which has application to many fields besides astrophysics. Thus, a large number of methods have been developed, with many important contributions being made by astrophysicists. A complete review of all the methods is beyond the scope of this discussion. For solving the equations of compressible fluid dynamics, the most popular methods include

- finite-difference techniques (which require hyper-viscosity to smooth discontinuities),
- finite-volume methods (which often use a Riemann solver to compute upwind fluxes),
- operator split methods (which combine elements of both finite-differencing and finite-volume methods for different terms in the equations),
- central methods (which often use simple expressions for the fluxes, combined with high-order spatial interpolation), and
- particle methods such as smooth particle hydrodynamics (SPH, which integrates the motion of discrete particles to follow the flow).

A good technical review of many of these methods is given by LeVeque (2002). SPH is an example of a method developed largely to solve astrophysics problems, although many of the developments in other methods (e.g. the extension of finite-difference and finite-volume methods to magnetohydrodynamics (MHD) to include the effects of magnetic fields on the dynamics of the fluid) have also been motivated by astrophysics.

Relation to other Fields

Computational astrophysics is necessarily inter-disciplinary, involving aspects of not only astrophysics, but also numerical analysis and computer science.

Numerical Analysis

Numerical Analysis is a rigorous branch of mathematics concerned with the approximation of functions and integrals, and the approximation of solutions to algebraic, differential, and integral equations. It provides tools to analyze errors that arise from the approximations themselves (truncation error), and from the use of finite-precision arithmetic on a computer (round-off error). Convergence, consistency, and stability of numerical algorithms are all essential for their use in practical applications. Thus, the development of new numerical algorithms to solve problems in astrophysics is deeply rooted in the tools of numerical analysis.

Computer Science

Computational science and computer science differ. The former is the use of numerical methods to solve scientific problems. The latter is the study of computers and computation. Thus, computer science tends to be more orientated towards the theory of computers and computation, while scientific computation is concerned with the practical aspects of solving science problems. Still, there is a large degree of overlap between the fields, with many computer scientists engaged in developing tools for scientific computation, and many scientists working on software problems of interest to computer scientists. Examples of the overlap include the development of standards for parallel processing such as the Message Passing Interface (MPI) and OpenMP, and development of parallel I/O filesystems such as Lustre.

Nuclear Astrophysics

Nuclear Astrophysics is a field at the intersection of nuclear physics and astrophysics that seeks to understand how nuclear processes shape the cosmos. In essence we look for the connection between properties of atomic nuclei and the properties of planets, stars, and galaxies. Open questions include "How did the universe create the elements?", "How can extremely dense and hot astrophysical environments be used to learn about fundamental properties of matter?", and "How is the energy created that powers stars and stellar explosions?". One fascinating aspect of this field is its interdisciplinarity and diversity. Work in nuclear astrophysics includes astronomical observations using telescopes, gravitational wave detectors, and neutrino detectors; accelerator laboratory experiments using beams of stable nuclei, radioactive nuclei, neutrons, and gamma-rays; laboratory analysis of interstellar grains; large scale computer simulations of stellar explosions and nuclei; and theoretical work in nuclear physics and astrophysics.

Nuclear Physics Enables Astrophysical Model Validation

Progress in nuclear astrophysics requires the development of hypotheses and models that describe what happens deep inside stars and stellar explosions. These models must then be compared with data to determine whether they indeed reflect what happens in nature. Without such model validation, progress is not possible. The challenge is that data on stellar interiors cannot be obtained directly. Rather, in most cases only a limited amount of radiation from the stellar surfaces or the debris of the explosion can be measured (exceptions include the detection of neutrinos from nearby supernovae or new data on stellar oscillations that can be used for Asteroseismology). Nuclear physics opens a unique validation path for astrophysical models by linking such observables to the conditions in the deep interiors. This is done by exploiting the strong dependence of nuclear processes on temperature, density, and composition, but requires accurate nuclear data. Once the nuclear physics is fixed, observables related to nuclear processes can be used to validate or falsify stellar models. In the future, model validation will become more important because of the increased complexity of stellar models, for example by moving from 1D to 3D calculations, or by including rotation. With more degrees of freedom in stellar models, validation will be key to ensure models reflect in a reasonable way what is actually happening inside a star or a stellar explosion.

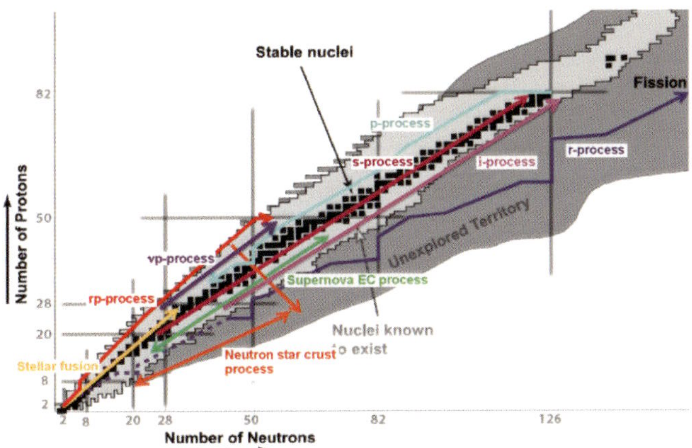

Figure: Schematic overview of the nuclear processes in the universe on the chart of nuclides

Models for Neutron Capture Processes

Validation through observed abundance patterns are of special importance for the neutron capture processes that produce elements beyond iron. These include the slow neutron capture process (s-process), the rapid neutron capture process (r-process), and possibly an intermediate neutron capture process (i-process). For these processes, detailed abundance patterns can be obtained from the solar system composition by de-convolution of multiple processes (s- and r-process), from abundance patterns in very metal poor stars that preserve signatures of neutron capture processes in the early Galaxy that polluted the material the star formed from (s, i, and r-process), and from meteoritic pre-solar grains (s-process, some r-process, novae, supernovae). These pre-solar grains were formed less than a billion years prior to solar system formation in stellar environments of the solar neighbourhood, and found their way into the solar system, where they can now be analyzed in the laboratory.

s-process: The site of the s-process, which produces about half of the heavy elements beyond germanium by a sequence of slow neutron captures and much faster β-decays, is known since the first observation of technetium on the surface of a red giant star in 1954. Technetium is an element that does not have a stable isotope, and must therefore be created inside the star it is observed in. Based on nuclear physics arguments, the only plausible way to produce the neutrons for an s-process in such a stellar environment is via mixing between hydrogen burning and helium burning layers. Hydrogen burning layers contain protons and 14N from hydrogen burning via the CNO cycle. When mixed into the helium burning layer, which contains helium and the helium burning product 12C, proton capture on 12C and the subsequent β-decay of 13N produce 13C, while helium capture reactions and β-decays on 14N produce 22Ne. When immersed in helium rich environments, both isotopes act as strong neutron sources via (α,n) reactions. S-process abundance observations are therefore intimately linked to mixing processes in stars, and can be used to directly probe quantitatively the mixing deep inside red giant stars. Uncertainties in the 13C(α,n) and 22Ne(α,n) reaction rates, as well as in the competing α capture rates, other potential neutron sources, and neutron absorbing reactions remain the main limitations of this approach and will have to be addressed in the future.

The s-process abundances itself also contain important information about the astrophysical environment. Neutron capture reactions in the main s-process are mostly in steady flow such that $Y < \sigma v >$ is constant (with Y being the produced s-process abundance of an isotope along the reaction path, and $< \sigma v >$ is the stellar reaction rate). This means that the nuclear reaction rate $< \sigma v >$ has a strong influence on the abundances produced, but that the sensitivity on the astrophysical environment is limited as neutron capture rates depend only weakly on temperature, and because the reaction path is fixed to the valley of stability. However, the steady flow equilibrium breaks down when the reaction sequence splits into parallel branches. At these branch points neutron captures and β-decays occur with comparable rates. Isotopic abundance ratios along different branches therefore become very sensitive to neutron density and, through the temperature sensitivity of the β-decay, on temperature. The experimental and theoretical determination of some of the key branch point rates enabled the validation of the today accepted multi-exposure s-process model in red giant (AGB) stars, where mixing processes activate both, the 13C(α,n) neutron source between thermal pulses, and the 22Ne(α,n) neutron source during helium flashes. With some experimental reaction rate data in place, branch point abundances could only be explained with both neutron exposures, and not with a single, average, neutron exposure, providing strong validation of the stellar models.

The s-process branch points occur per definition at unstable nuclei. Measurements of neutron capture rates on these isotopes are difficult and many of the shorter-lived cases have not been determined experimentally. To maximize the validation potential of stellar models through s-process nucleosynthesis, measurements of neutron capture rates on these radioactive species are important for the future. In some cases, samples for such measurements could be provided by new radioactive beam facilities using isotope harvesting techniques, but depending on the half-lives this requires a nearby, or in house, neutron beam facility.

r-process: The r-process is the second major nucleosynthesis process producing elements beyond germanium . It proceeds via rapid neutron captures and β-decays, possibly including (γ,n) photodsintegration reactions and fission processes. Unlike for the s-process, the r-process site has not been determined with certainty. A large number of hypothetical models are therefore discussed in the literature. Among the more recently proposed sites are direct ejecta from neutron star mergers, neutrino driven winds in neutron star mergers or core collapse supernovae, magneto-hydrodynamically driven jets in rapidly rotating supernovae, or a neutrino driven r-process in He shells of supernovae. In this situation, nuclear physics provides a possible pathway to validate models through comparison with observed r-process elemental abundances. The elemental abundance patterns produced by the r-process are well known from observations of metal-poor stars that preserve the composition of r-process events in the early Galaxy. Observational constraints may in the future distinguish between supernovae and neutron star mergers, and first hints in favor of neutron star mergers are being obtained from the observations of optical afterglows of short γ-ray bursts, so called kilo-novae and the r-process enrichments observed in some dwarf galaxies. However, nuclear physics seems to be the only way to validate individual sites and mechanisms within these broad scenarios. Unlike the s-process, the r-process reaction sequence includes very neutron-rich unstable isotopes. In low temperature scenarios, the so called cold r-process, neutron capture rates are comparable to β-decay rates along the r-process path. In the hot r-process, the path is determined by equilibrium between neutron capture and photo-disintegration. In either case, the path strongly depends on neutron density, in the hot case also on temperature, and is expected to change dramatically during the r-process. As the path sweeps over the chart of nuclides, characteristic abundance patterns are formed depending on the nuclear properties such as β-decay half-lives, β-delayed neutron emission probabilities, neutron capture rates (in the cold r-process case) and nuclear masses (in the hot r-process case) along a the momentary reaction sequence. As the r-process reaches the actinide region and beyond, fission also plays a critical role. Once the nuclear physics is known, this superposition of characteristic abundance patterns can be calculated for a given astrophysical model and then be compared with observations to distinguish between the proposed astrophysical environments and astrophysical assumptions. The goal is not to reproduce the observed abundances accurately (which requires to address nuclear and astrophysical uncertainties, with the latter often overwhelmingly large) but, by fixing the nuclear physics, to create validation pathways and guidance for the astrophysical aspect of the problem. A potential difficulty are freeze out effects. As the neutron density drops and the r-process ends some abundance features may be washed out, and new ones can be created for example through late time fission or preferential capture of neutrons in certain regions. Constraints on the freeze out behavior may also provide valuable validation opportunities, though.

The long standing challenge for the field has been to produce the extremely unstable nuclei along the r-process path to enable measurements of the critical nuclear physics quantities. This is one of the primary motivations for a next generation of radioactive beam facilities such as RIBF, FAIR, and FRIB. Despite of these difficulties, some experiments have succeeded to reach the r-process - examples include the half-life measurement of ^{130}Cd at ISOLDE/CERN , the half-life measurement of ^{78}Ni at NSCL/MSU, mass measurements at ATLAS/ANL , recent measurements near N = 126 at FRS/GSI , and the first results from a next generation radioactive beam facility at RIBF/RIKEN where a broad range of new half-lives in the r-process has been determined. A similar challenge has been the theoretical prediction of nuclear properties along the r-process path with sufficient precision so as to be useful for astrophysical applications. R-process nuclei are heavy and in general, with a few exceptions near closed shells, well beyond the reach of large scale shell model calculations. While a number of very useful global models for the prediction of masses, decay properties, and reaction rates, exist, their accuracy is not sufficient for astrophysical applications and their uncertainty is difficult to be characterized in the absence of nuclear data. Indeed, differences in predictions of the various models indicate much larger uncertainties than a comparison with data near stability would imply. There are however a number of very promising theoretical developments, including Density Functional Theory approaches, which also provide estimates for uncertainties. Examples include predicted masses or fission processes. To be useful for astrophysical applications it will be important to create global data sets with theoretical predictions for all nuclei (the entire mass and element range, and also odd-even and odd-odd nuclei), even if that means to compromise the quality of the prediction in some areas. This will be essential to ensure progress in nuclear theory translates into progress in nuclear astrophysics.

i-process: Recently, an intermediate neutron capture process (i-process) with neutron capture timescales in between the s- and the r-processes has been proposed to occur in red giant stars, especially in old stars with low metal content, when hydrogen and helium burning zones mix more easily, leading to entrainment of hydrogen into helium burning regions. This has been motivated by the identification of these entrainment processes in 3D fluid dynamics calculations. Validation of such new and surprising results requires the use of i-process nuclear data to predict the characteristic abundance signatures in these new astrophysical models. These signatures have to then be matched to observed signatures that cannot be explained with other processes. Two of such (potential) signatures have been identified: (1) The peculiar abundances measured on the surface of Sakurai's object, a star at the very end of its red-giant phase that is thought to have been already on a path towards becoming a planetary nebula when one more thermal pulse ignited the star for one last time. (2) The abundances detected in a class of metal poor stars with strong carbon enhancement that exhibit heavy neutron capture elements that neither fit a solar pattern, an r-process pattern, or a s-process pattern. The standard explanation has been to invoke a mix of s- and r-processes, but at least in some cases it has been difficult to obtain a good fit of the observations with this approach. A recently proposed alternative explanation is that some or all of these stars may have been polluted by an i-process in the early Galaxy.

While still somewhat speculative, it is very important for nuclear physics to contribute reliable data for i-process models so that models can be validated (or falsified) in the context of the observational signatures. This is the only way to determine whether the i-process indeed is responsible for the observed abundance patterns or not. In terms of nuclear physics needs, the i-process is unique in that it requires knowledge of neutron capture rates on unstable neutron deficient nuclei a few mass units away from stability. Determining neutron capture rates on unstable nuclei has been a formidable challenge for experimental and theoretical nuclear physics, because both, neutron and target nucleus are unstable and traditional experiments with a stable or unstable beam impinging on a stable target are not viable. A variety of techniques have been developed, but all have strong limitations and significant systematic uncertainties. These include experiments that inform input to statistical reaction rate calculations, such as measurements of low lying dipole strength or the new beta-Oslo method, transfer reactions, Coulomb breakup, or other so called surrogate techniques. The i-process problem, and the fact that most of the important reactions involve nuclei that are not too exotic and can be produced in significant quantities at current facilities, provides a strong motivation to push the development of these techniques now.

Neutrinos

With the detection of solar neutrinos in the 1960s and neutrinos from supernova 1987A, a new era of neutrino astronomy has begun. As neutrinos are messengers that probe the deepest interiors of stars and supernova explosions, their detection opens up exciting opportunities to directly validate models of stellar environments not easily accessible by other observational probes. Nuclear physics is key in this validation process as the neutrinos are produced by nuclear processes - the pp-chains and the CNO cycle in the Sun, and electron capture reactions in the collapsing core of a supernova. The validation of the standard solar model through the detection of solar neutrinos is one of the prime examples that illustrates the importance of accurate nuclear physics for validating stellar models. It was only through decade long work in nuclear theory and experiment that the nuclear reactions in the core of the sun were well enough understood to be able to demonstrate that the detected neutrino flux was lower than expected, the so called solar neutrino problem. This ultimately led, together with other observations, to the conclusive discovery of neutrino oscillations and, once the properties of neutrinos were understood, to a convincing validation of our understanding of the conditions in the centre of the Sun.

The detection of neutrinos from a Galactic supernova would open up similar opportunities for the validation of core collapse supernova models, especially the explosion mechanism. Unfortunately these events are rare (a few per century at most) and so far none has occurred since neutrino detectors became operational. Current neutrino detectors would detect 1000s of neutrinos from such an event. An important goal to enable this validation pathway is to determine with sufficient precision the nuclear reactions that drive the neutrino emission and shape the expected neutrino signal. The relevant reactions have been recently identified and experimental and theoretical work is ongoing to address this nuclear physics need. One experimental approach is the use of charge exchange reactions that can simulate weak interaction processes and can be used to extract the relevant transition rates.

P-process Models

The p-process is responsible for the origin of 35 stable isotopes that are too neutron deficient to be produced by neutron capture processes. Yet, nature has found a path to create these isotopes, albeit with typically factors of 10-100 lower abundance than other more neutron rich isotopes of the same element. A number of sites have been proposed, including the outer layers of various types of type Ia supernovae (white dwarf mergers, accreting white dwarfs that ignite because their mass exceeds the Chandrasekhar mass limit, or white dwarf explosions triggered by the ignition of an accreted helium layer) or the shock front passing through the outer layers of a star undergoing a core collapse supernova explosion. The common feature of most current p-process scenarios is that the p-nuclei are produced by photo-disintegration of a distribution of stable seed nuclei. These seed nuclei were produced by previous stellar generations, by an s-process in the exploding star or, in the case of accreting scenarios, in the companion star from which matter is transferred from.

A challenge for current attempts to validate p-process models is that neither very many observations are available, nor the nuclear physics is well understood. The observational challenge is that stellar spectroscopy is insensitive to the minute amounts of p-isotopes contributing the the abundance of a particular element. This has traditionally left the solar abundance distribution as the only reference, which may be a superposition of multiple p-processes and may therefore be only of limited use to guide the identification of the site. Meteoritic abundances may provide the only other isotopic information and are therefore very valuable. For example, the isotopic composition of some meteorites provides strong evidence for the presence of long lived, radioactive p-process isotopes in the early solar system such as 92Nb and 146Sm, which have since decayed. On the nuclear physics side, key reactions have been identified for a core collapse supernova scenario and involve a large number of stable and unstable nuclides. While impressive progress has been made in measuring p-process reactions on stable nuclei , the reactions on unstable isotopes remain uncertain.

Because of these challenges, and because the p-process is rather robust owing to the small amount of material that needs to be produced, validation of p-process models has so far not allowed researchers to discriminate among a large range of models. Rather, many p-process models can reproduce most p-process abundances within typically a factor of 3. An exception are the light p-nuclides $^{92,94}Mo$ and $^{96,98}Ru$ which have unusually large isotopic abundances and are underproduced by most pprocess models. An interesting aspect of the p-process is that it can be used to not only validate the temperature conditions in the explosive nucleosynthesis site, but also the initial composition of the seed nuclei. Indeed, recent models of the p-process in type Ia supernova have attempted to address the underproduction of $^{92,94}Mo$ and $^{96,98}Ru$ by assuming an enhancement in s-process elements in the companions star that triggers the explosion. This has only been partially successful as even with favorable parameter choices the large abundance of 94Mo remains a puzzle. The alternative of a rapid proton capture process or a vp-process serving as additional source of the enhancement of the light p-nuclei is disfavored by the similarly large enhancement of radioactive ^{92}Nb, which is shielded from these processes but can be readily produced in a traditional photodisintegration based p-process.

As radioactive beam facilities in the future can address the remaining nuclear physics uncertainties in the p-process it will be interesting to see whether model discrimination becomes possible by comparing to solar and meteoritic p-process abundances. A promising step in this direction is recent work that demonstrated that the p-process synthesis of 94Mo occurs in a rather narrow temperature range. Once the nuclear physics is fully understood, the known natural abundance of this isotope can then be used to validate specific aspects of the conditions in the explosion. It would be interesting to systematically investigate the sensitivity of various p-isotopes to temperature and density conditions, and to specific features in the distribution of seed nuclei, and identify the ones with the most power towards model discrimination. Another focus should be the nuclear uncertainties affecting the production of the long lived radioactive p-isotopes 92Nb and 146Sm, which need to be minimized to use meteoritic data for validation. Future nuclear physics studies could then concentrate on the relevant reaction pathways that produce these key isotopes.

Nova Models

Nova explosions occur on the surface of white dwarfs that transfer matter onto their surface from a companion star in a stellar binary system. The nova explosion leads to a dramatic brightening of the star for many months. It is powered by proton capture reactions on stable and unstable neutron deficient isotopes. Validating nova models is important to understand the evolution of accreting white dwarfs in general, especially as such systems (not necessarily the ones that exhibit novae though) may be a pathway to explain type Ia supernova explosions. Compared to other astrophysical sites, nova models are in a somewhat curious position in terms of model validation. Despite of the strong sensitivity on difficult to determine nuclear reactions on unstable nuclei, the nuclear physics of nova explosions is relatively well understood thanks to decades of measurements at stable and radioactive beam facilities. The main reason is that the unstable nuclei are relatively close to stability and can be produced with sufficient intensity, and that temperatures during the explosion are high, resulting in much larger and easier to measure reaction rates compared to other stellar environments. Nevertheless, some key reactions in the most extreme class of novae, explosions on the surface of more massive oxygen-neon-magnesium white dwarfs, remain to be determined, especially in the context of the synthesis of heavier elements such as sulfur. An example is the rate of proton capture on unstable ^{30}P, which serves as a bottleneck for the synthesis of heavier elements and needs to be known to use observations of sulfur abundances for model validation. As the development of radioactive beams progresses with new techniques and facilities, these remaining measurements should become feasible soon. It is possible that there is a subset of rare nova events where accretion rates and white dwarf temperatures lead to the accumulation of significantly more matter, with more extreme explosions and significantly extended nucleosynthesis. In such events, a much broader range of nuclear reactions would become important to understand in order to predict observables that could be used to identify such events.

The sensitivity of the observed composition of nova ejecta to conditions during the explosion has been investigated in detail, paving the way for using these observations as, for example, temperature probes of the exploding layer. This delineates a clear pathway for how model validation can be performed. However, a major challenge in the context of nova model validations are limitations in observational data. The number of novae with accurately determined abundance patterns that cover a broad range of elements is small. One issue is the limited availability of space based observatories with UV spectroscopy capability. Furthermore, such composition measurements are difficult to interpret and have large uncertainties, because of the rapidly evolving and complex structured ejecta cloud. This could be circumvented by detecting the γ-radiation of radioactive isotopes produced in the explosion, which penetrates more easily the ejected material. There are a number of suitable isotopes such as ^{22}Na, or, in the early stages of the explosion, ^{18}F. However, current γray observational capabilities in the MeV range are not sufficient to make meaningful measurements of novae at typical distances. The analysis of pre-solar grains that are embedded in meteorites and may originate from dust formed in the ejecta of nearby novae provides a tantalizing opportunity to perform more detailed composition measurements. However, the nova origin of a particular grain is difficult to establish - more precise nuclear physics may help guide this identification process through comparison with the expected isotopic abundance pattern. Nevertheless, one has to be careful that this does not become a circular argument, where nova models are used to identify nova grains, which are then used to validate the nova model.

Other Theoretically Predicted Nucleosynthesis Sites

Theoretical models sometimes predict novel sites of nucleosynthesis. Whether these predicted processes, or their sites, exists at all must then be determined empirically. Nuclear physics is needed to make accurate predictions of the characteristic patterns of elemental or isotopic abundances produced in these hypothetical sites that one can then search for in observations. An example are so called Thorne-Zytkow objects - ˙ red giant stars with a neutron star core where a pulsed or interrupted rapid proton capture process may produce anomalous abundance patterns of Rb, Sr, Zr, and Mo. While theoretical models indicate that stable Thorne-Zytkow objects cannot ˙ form, the discovery of stars with the predicted characteristic signatures have been recently reported. In addition, it may be that even a temporary formation of a Thorne-Zytkow like object during a merger of a star and a neutron star may produce ˙ interesting abundance signatures. It would therefore be useful to take advantage of recent improvements in our understanding of the nuclear physics of the rp-process to improve predictions of the abundance signatures produced by Thorne-Zytkow objects. ˙ Other examples for theoretically predicted nucleosynthesis sites include the neutrino-p process in proton rich neutrino driven winds emerging from the nascent neutron star in core collapse supernovae or the extreme nova explosions predicted to occur on particularly cold accreting white dwarf stars. Also in these cases, the search for the signatures of these events has to be guided by predictions based on reliable nuclear physics.

Chemical Evolution Models

There has been tremendous progress in astronomical observations of the compositions of stars. As the number of stellar composition data continues to grow dramatically, astrophysical models of single nucleosynthesis sources are not sufficient anymore to link nuclear physics with these observations and to take advantage of the wealth of additional information on abundance variations and correlations. Each star's composition reflects the composition of the Galaxy at the time and location of its formation (for the subset of stars were the composition has not been modified at a later stage, for example through mass transfer in a binary system). This composition is determined not only by a multitude of nucleosynthesis processes, but also by how ejecta from nucleosynthesis sites are mixed over time, which in turn depends on the complete merger history of the various stellar components that formed our Galaxy. Models of individual nucleosynthesis events must therefore be integrated into a model of Galactic chemical evolution to enable comparison, in a statistical sense, to the complete set of observational data. There has been significant progress recently in developing chemical evolution models based on N-body simulations of Galaxy formation.

Improved nuclear physics in connection with nucleosynthesis calculations for a wide range of stellar masses and metallicities (so called "yield grids") in connection with the new observational datasets offers the opportunity to validate these models, our understanding of chemical evolution, and come to a true understanding of the origin of the elements, one of the ultimate goals of this field. This significantly broadens the scope of nuclear astrophysics, which becomes a tool to address questions of Galaxy formation, star formation, and galactic flows of matter. Instead of a simple comparison of abundance patterns, this validation now also considers the frequency of the nucleosynthesis event in question, how it depends on stellar parameters, and how the ejected abundance patterns vary over the history of the Galaxy. Such validation will be key for identifying convincingly sites of nucleosynthesis such as the r-process site.

One may argue that the large uncertainties in chemical evolution models and the large number of unknown parameters, that are then convoluted with the considerable uncertainties in the stellar nucleosynthesis sources, make it impossible to draw meaningful conclusions. Recent work has demonstrated that this is not necessarily so. Rather, the large number of observable elements, each probing different types of environments and sites, allows to control some of the parameter degeneracy. For example, a recent analysis demonstrated that deficiencies in stellar yield grids, and therefore in stellar models, can be identified in chemical evolution models even when taking into account the uncertainties of the chemical evolution process.

Instead of looking at the chemical evolution of an entire galaxy, it is more straight forward to investigate smaller, to some extent self-contained, sub systems such as Globular Clusters or Dwarf Galaxies. An example is recent work that took advantage of experimental progress in reducing the nuclear physics uncertainties in stellar reaction rates achieved at stable beam facilities to explain the anomalous abundance trends observed in the globular cluster NGC2419.

Viewing the Invisible - Probing Stars with Nuclear Physics

Closely related to model validation is the extraction of parameters of a stellar system using precision nuclear physics and observations. This works best for models that have been validated for cases where all parameters are known, and are then applied to unknown systems. In this approach, nuclear physics essentially leads to a dramatic enhancement of the insights that can be gained from observations.

Probing the First Stars

One of the most exciting examples of this are attempts to observe, characterize, and understand the very first stars that formed from the chemically pristine material created by the Big Bang. While advanced infra-red telescopes such as the planned James-Webb Space Telescope attempt to provide this information by pushing observations to the faintest and most distant (and red-shifted) objects, observations of the composition of nearby stars in our Galaxy in combination with nuclear physics open an alternative pathway. This approach has been termed "Near Field Cosmology" because of the constraints that can be obtained on the environment in the early universe. Indeed it has been proposed that a class of rare carbon enhanced metal poor (CEMP) stars without detectable levels of neutron capture elements (CEMPno) may have formed from the debris of the first supernova explosions after the Big Bang. With accurate nuclear physics, the observed composition can be matched to supernova models, where parameters such as the initial mass of the exploding star and the explosion energy can be varied to extract some of the properties of these early stars. An example is SMSS J031300.36-670839.3 where in addition to carbon also magnesium and calcium have been detected. Based on the pattern of the detected chemical abundances it was suggested that a relatively weak explosion of a first star with an initial mass of 60 solar masses created the observed signature. However, the nuclear processes responsible for the calcium production should be investigated and nuclear uncertainties addressed. In addition to hot hydrogen burning it was also proposed that a mild rp-process may occur in such massive early stars, albeit with larger mass.

Probing the Sun

A closer to home example for how nuclear physics can be used to peek inside a star are plans to determine the metal content in the interior of the Sun. This can be done by using precise nuclear rates for the reactions in the CNO cycle and observations of neutrinos emitted from these reactions deep in the core of the Sun. With the well known solar temperature and density profiles, the amount of CNO nuclei can then be determined. This is interesting because the composition inferred from absorption spectra from the solar photosphere using sophisticated 3D atmosphere models is in disagreement with helioseismic data. Neutrino experiments such as BOREXINO are improving their backgrounds with the goal to achieve a first measurement of the flux of CNO neutrinos from the Sun. At the same time, work is underway to determine with sufficient precision the proton capture rates in the CNO cycle. The determination of such stellar reaction rates has been a major challenge for the field because of the very small rates. So far, with very few exceptions, none of the nuclear reactions in stars have been measured at the relevant energies, rather experiments are performed at higher energies and theory is used to extrapolate. This challenge is now being addressed with new technical developments, including the installation of high intensity accelerators underground.

Probing Neutron Stars

Neutron stars are very special objects that truly straddle the boundary between nuclear physics and astrophysics. As km sized blobs of nuclear matter, they squarely fall into the realm of nuclear physics, though as astrophysical objects in space they are studied by astronomers with astronomical techniques instead of accelerators and detectors. The determination of neutron star properties from observations therefore not only provides information on stellar physics, but on the fundamental properties of nuclear matter. The recent discovery of a neutron star with a mass of about two solar masses immediately ruled out large classes of possible equations of state for nuclear matter because it requires that compressed nuclear matter can provide enough pressure to balance the gravity for such a massive object.

Accreting neutron stars in stellar binary system accumulate matter that is transferred from a regular companion star. The mass transfer onto the neutron star creates a wealth of observables which can be linked through accurate nuclear physics data to neutron star and nuclear matter properties. As the mass transfer rate varies, and sometimes stops entirely, the response of the neutron star can be observed, making accreting neutron stars a unique laboratory to study dense states of matter.

One key observable are type I x-ray bursts, thermonuclear explosions of the matter that accumulates on the neutron star surface that repeat within hours or days and are the most frequent thermonuclear explosions observed in astrophysics. X-ray burst light curves and the spectral evolution during X-ray bursts has been used to constrain masses and radii of the underlying neutron star but the large range of unknown system parameters such as the composition of the accreted matter are a problem. It turns out though that the 10-100 s long X-ray light curve is sensitive to these parameters. The light curves are powered by the αp- and rapid proton capture processes (rp-process). Once the nuclear physics is sufficiently well understood, models of burst light curves can be compared to observations of a particular source to constrain accretion rate, accreted composition, peak luminosity (and distance), and thermal conditions of the neutron star surface. With these constraints, neutron star properties can then be extracted, and other open questions can be addressed. With still higher nuclear precision it may even be possible to use the shape of the light curve directly to probe gravity on the neutron star surface.

The observation of the elements produced in X-ray burst nuclear reactions would open another very powerful pathway to understand accreting neutron stars. In particular, the redshift of identified spectral features created by these elements would provide direct constraints on the neutron star compactness. So far, no clear elemental signatures have been extracted from X-ray spectra - an example for a recent tentative measurement is the report of a possible chromium feature observed in GRS 1741.9- 2853 by the NUSTAR satellite. On the other hand, theoretical studies indicate that mass ejection of burst ashes via stellar winds should be expected for a subset of particularly powerful bursts. What is ejected is not the final burst ashes, but the intermediate composition created at the beginning of the burst at the time where the convection zone has its maximum extent (unless subsequent bursts are mixing up material from previous bursts). While frequency and ejected mass of such bursts are probably too low for a significant contribution to the elements found in the solar system, the resulting spectral features are predicted to be strong enough to be detectable with current instruments. It will be important to continue the search for spectral features in X-ray bursts, and theoretical work on predicting the expected signatures.

Great strides have been made by the nuclear physics community to determine the nuclear reactions on very neutron deficient radioactive isotopes in the rp- and αpprocesses. Masses and decay properties are key quantities that have mostly been determined by measurements at radioactive beam facilities over the last decades. However, the remaining challenge for the future are uncertainties in proton and α-capture rates.

Another important observable for accreting neutron stars is the long term cooling of the crust that was heated by nuclear reactions during accretion. This cooling can be observed over many years in transient accretors where the accretion shuts off for years or decades. A dozen or so systems have been observed so far. The cooling behavior probes the physics of dense matter in the neutron star crust and may be sensitive to various phase transitions, including the presence of superfluid neutrons and nuclear pasta. However the quantitative comparison of model curves with observations to extract this information requires a good understanding of the electron capture, β-decay, and fusion reactions of very neutron rich rare isotopes that heat and cool the neutron star crust during the accretion phase. All the critical nuclei become now accessible at the new generation of radioactive beam facilities and experiments geared towards addressing the nuclear physics questions related to neutron star crusts are an important new science motivation. Which reactions are the important ones depends also sensitively on the composition created by the X-ray bursts during the accretion phase. These burst ashes replace the original neutron star crust and set its composition. The nuclear physics of X-ray bursts is therefore another important element in the context of neutron star crust studies. A possible strategy could be to use precise αp- and rp-process nuclear physics to fit burst models to the observed burst light curves of a particular transiently accreting system to predict that system's crust composition. This information can then be combined with experimental and theoretical nuclear data on extremely neutron rich nuclei to calculate crust cooling models, that can then be compared to observations during the accretion-off state to probe the dense matter inside the neutron star crust.

Nuclear Physics Enables Nucleosynthesis Predictions

Beyond model validation and the probing of stellar environments, nuclear physics is key to predict individual nucleosynthesis contributions to elements and isotopes for which its not possible to disentangle processes observationally. In this case, models that have been validated otherwise (maybe with different processes or using different elemental abundance ranges) have to be outfitted with precise nuclear physics to predict the possible abundance pattern or patterns that a particular process may contribute. These predictions can then be used to disentangle the contributions to observed abundance patterns using chemical evolution models that also take into account the frequency, spatial distribution, and amount of ejected material of the various different stellar nucleosynthesis sources, as well as the time and space variability of these quantities.

An important example are the"neutron capture" elements from germanium to about palladium. Observations of abundance patterns in metal poor stars clearly indicate that multiple processes contribute in this element range, and that the observed patterns are to some extent random mixtures of these processes. The processes that potentially contribute in this element range may include the classical weak s-, and main s-processes, a main r-process, a weak r-process, an explosive charged particle process, the neutrino-p process in core collapse supernovae, or the Light Element Primary Process (LEPP) invoked to explain deficiencies in standard s-process models. These processes span stable, neutron rich, and neutron deficient isotopes and a wide range of nuclear reaction types. The relevant reaction paths are not as far from stability as for the r-process producing heavier elements, and fission processes from the main r-process are not expected to contribute strongly. This makes the nuclear physics of the synthesis of elements in the germanium to palladium range an important target for current and future stable, neutron beam, and radioactive beam facilities where it should be possible to obtain complete nuclear data sets that will enable an understanding of the origin of these elements.

Another example is the prediction of the nucleosynthesis contribution to individual isotopes. In most cases, stellar observations are limited to elemental abundances and can therefore not be used to disentangle contributions to individual isotopes. An example are the s- and r-processes, where the main source of information on isotopic abundance patterns comes from the solar system composition, where both processes are mixed. While so called s- or r-only isotopes (isotopes that can only be reached by one of the two processes) and observed elemental abundances can provide guidance and serve as model normalization points, isotopic abundances inbetween these data points have to be interpolated using nucleosynthesis models and accurate nuclear physics. In the past, high precision information on s-process neutron capture rates has been key in predicting the s-process contributions to isotopes beyond iron, enabling the extraction of the predicted isotopic r-process abundances. In the future it may become possible with accurate r-process nuclear physics from advanced radioactive beam facilities to do the same for the r-process. This would allow to search for additional nucleosynthesis contributions beyond the s- and r-processes.

References

- Astrophysics: thecanadianencyclopedia.ca, Retrieved 21 April 2018
- Solar-physics-204: chegg.com, Retrieved 31 March 2018
- Astrophysics-Observational-astrophysics: newworldencyclopedia.org, Retrieved 11 July 2018
- Astrophysics-Theoretical-astrophysics: newworldencyclopedia.org, Retrieved 31 June 2018
- Computational-astrophysics: scholarpedia.org, Retrieved 30 March 2018
- What-is-nuclear-astrophysics: jphysplus.iop.org, Retrieved 20 May 2018

2

Orbital Mechanics

The field of study concerned with the application of celestial mechanics and ballistics to the motion of rockets and spacecraft is known as orbital mechanics. Some of the topics central to the development of orbital mechanics such as orbital velocity, Kepler's laws of planetary motion, gravity assist, Hohmann transfer orbit, geostationary transfer orbit, etc. have been covered in extensive detail in this chapter.

The path of Earth satellite revolving around the earth is known as orbit. This path can be represented with mathematical notations. Orbital mechanics is the study of the motion of the Earth satellites that are present in orbits.

Orbital Elements

Orbital elements are the parameters, which are helpful for describing the orbital motion of satellites. Following are the orbital elements.

- Semi major axis
- Eccentricity
- Mean anomaly
- Argument of perigee
- Inclination
- Right ascension of ascending node

The above six orbital elements define the orbit of earth satellites. Therefore, it is easy to discriminate one satellite from other satellites based on the values of orbital elements.

Semi Major Axis

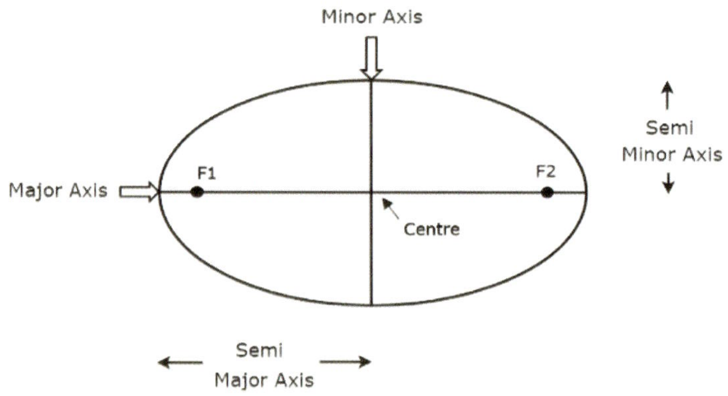

The length of Semi-major axis (a) defines the size of satellite's orbit. It is half of the major axis. This runs from the center through a focus to the edge of the ellipse. So, it is the radius of an orbit at the orbit's two most distant points.

Both semi major axis and semi minor axis are represented in above figure. Length of semi major axis (a) not only determines the size of satellite's orbit, but also the time period of revolution.

If circular orbit is considered as a special case, then the length of semi-major axis will be equal to radius of that circular orbit.

Eccentricity

The value of Eccentricity (e) fixes the shape of satellite's orbit. This parameter indicates the deviation of the orbit's shape from a perfect circle.

If the lengths of semi major axis and semi minor axis of an elliptical orbit are a & b, then the mathematical expression for eccentricity (e) will be

$$e = \frac{\sqrt{a^2 - b^2}}{a}$$

The value of eccentricity of a circular orbit is zero, since both a & b are equal. Whereas, the value of eccentricity of an elliptical orbit lies between zero and one.

The following figure shows the various satellite orbits for different eccentricity (e) values

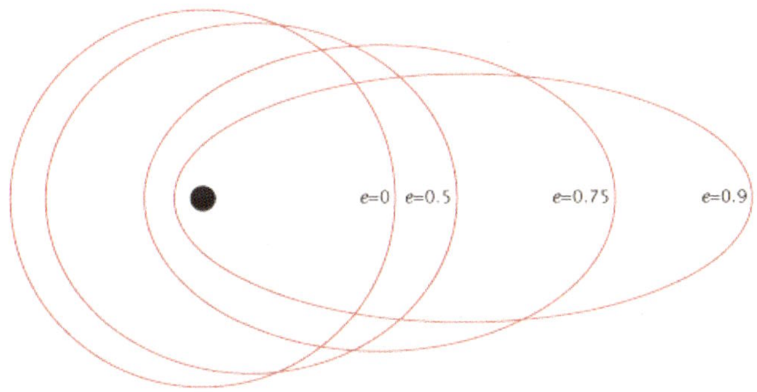

In above figure, the satellite orbit corresponding to eccentricity (e) value of zero is a circular orbit. And, the remaining three satellite orbits are of elliptical corresponding to the eccentricity (e) values 0.5, 0.75 and 0.9.

Mean Anomaly

For a satellite, the point which is closest from the Earth is known as Perigee. Mean anomaly (M) gives the average value of the angular position of the satellite with reference to perigee.

If the orbit is circular, then Mean anomaly gives the angular position of the satellite in the orbit. But, if the orbit is elliptical, then calculation of exact position is very difficult. At that time, Mean anomaly is used as an intermediate step.

Argument of Perigee

Satellite orbit cuts the equatorial plane at two points. First point is called as descending node, where the satellite passes from the northern hemisphere to the southern hemisphere. Second point is called as ascending node, where the satellite passes from the southern hemisphere to the northern hemisphere.

Argument of perigee (ω) is the angle between ascending node and perigee. If both perigee and ascending node are existing at same point, then the argument of perigee will be zero degrees

Argument of perigee is measured in the orbital plane at earth's center in the direction of satellite motion.

Inclination

The angle between orbital plane and earth's equatorial plane is known as inclination (i). It is measured at the ascending node with direction being east to north. So, inclination defines the orientation of the orbit by considering the equator of earth as reference.

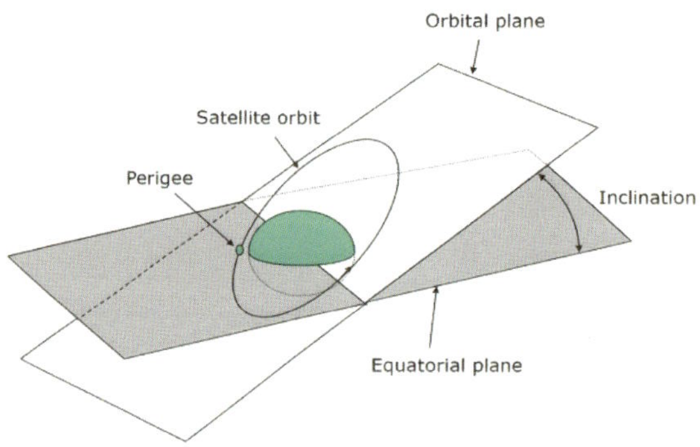

There are four types of orbits based on the angle of inclination.

- Equatorial orbit – Angle of inclination is either zero degrees or 180 degrees.
- Polar orbit – Angle of inclination is 90 degrees.
- Prograde orbit – Angle of inclination lies between zero and 90 degrees.
- Retrograde orbit – Angle of inclination lies between 90 and 180 degrees.

Right Ascension of Ascending node

We know that ascending node is the point, where the satellite crosses the equatorial plane while going from the southern hemisphere to the northern hemisphere.

Right Ascension of ascending node (Ω) is the angle between line of Aries and ascending node towards east direction in equatorial plane. Aries is also called as vernal and equinox.

Satellite's ground track is the path on the surface of the Earth, which lies exactly below its orbit. The ground track of a satellite can take a number of different forms depending on the values of the orbital elements.

Practical Techniques

Rules of Thumb

The following rules of thumb are useful for situations approximated by classical mechanics under the standard assumptions of astrodynamics outlined below the rules. The specific example discussed is of a satellite orbiting a planet, but the rules of thumb could also apply to other situations, such as orbits of small bodies around a star such as the Sun.

- Kepler's laws of planetary motion:
    - Orbits are elliptical, with the heavier body at one focus of the ellipse. Special case of this is a circular orbit (a circle is a special case of ellipse) with the planet at the center.
    - A line drawn from the planet to the satellite sweeps out *equal areas in equal times* no matter which portion of the orbit is measured.
    - The square of a satellite's orbital period is proportional to the cube of its average distance from the planet.
- Without applying force (such as firing a rocket engine), the period and shape of the satellite's orbit won't change.
- A satellite in a low orbit (or low part of an elliptical orbit) moves more quickly with respect to the surface of the planet than a satellite in a higher orbit (or a high part of an elliptical orbit), due to the stronger gravitational attraction closer to the planet.
- If thrust is applied at only one point in the satellite's orbit, it will return to that same point on each subsequent orbit, though the rest of its path will change. Thus one cannot move from one circular orbit to another with only one brief application of thrust.
- From a circular orbit, thrust applied in a direction opposite to the satellite's motion changes orbit to elliptical; the satellite will descend and reach the lowest orbital point (the periapse) at 180 degrees away from the firing point; then it will ascend back. Thrust applied in the direction of the satellite's motion creates an elliptical orbit with its highest point (apoapse) 180 degrees away from the firing point.

The consequences of the rules of orbital mechanics are sometimes counter-intuitive. For example, if two spacecraft are in the same circular orbit and wish to dock, unless they are very close, the trailing craft cannot simply fire its engines to go faster. This will change the shape of its orbit, causing it to gain altitude and actually slow down relative to the leading craft, missing the target. The space rendezvous before docking normally takes multiple precisely calculated engine firings in multiple orbital periods requiring hours or even days to complete.

To the degree that the standard assumptions of astrodynamics do not hold, actual trajectories will vary from those calculated. For example, simple atmospheric drag is another complicating factor for objects in low Earth orbit. These rules of thumb are decidedly inaccurate when describing two or more bodies of similar mass, such as a binary star system. Celestial mechanics uses more general rules applicable to a wider variety of situations. Kepler's laws of planetary motion, which can be mathematically derived from Newton's laws, hold strictly only in describing the motion of two gravitating bodies in the absence of non-gravitational forces; they also describe parabolic and hyperbolic trajectories. In the close proximity of large objects like stars the differences between classical mechanics and general relativity also become important.

Laws of Astrodynamics

The fundamental laws of astrodynamics are Newton's law of universal gravitation and Newton's laws of motion, while the fundamental mathematical tool is his differential calculus.

Every orbit and trajectory outside atmospheres is in principle reversible, i.e., in the space-time function the time is reversed. The velocities are reversed and the accelerations are the same, including those due to rocket bursts. Thus if a rocket burst is in the direction of the velocity, in the reversed case it is opposite to the velocity. Of course in the case of rocket bursts there is no full reversal of events, both ways the same delta-v is used and the same mass ratio applies.

Standard assumptions in astrodynamics include non-interference from outside bodies, negligible mass for one of the bodies, and negligible other forces (such as from the solar wind, atmospheric drag, etc.). More accurate calculations can be made without these simplifying assumptions, but they are more complicated. The increased accuracy often does not make enough of a difference in the calculation to be worthwhile.

Kepler's laws of planetary motion may be derived from Newton's laws, when it is assumed that the orbiting body is subject only to the gravitational force of the central attractor. When an engine thrust or propulsive force is present, Newton's laws still apply, but Kepler's laws are invalidated. When the thrust stops, the resulting orbit will be different but will once again be described by Kepler's laws. The three laws are:

- The orbit of every planet is an ellipse with the sun at one of the foci.
- A line joining a planet and the sun sweeps out equal areas during equal intervals of time.
- The squares of the orbital periods of planets are directly proportional to the cubes of the semi-major axis of the orbits.

Escape Velocity

The formula for an escape velocity is easily derived as follows. The specific energy (energy per unit mass) of any space vehicle is composed of two components, the specific potential energy and the specific kinetic energy. The specific potential energy associated with a planet of mass M is given by

$$\epsilon_p = -\frac{GM}{r}$$

while the specific kinetic energy of an object is given by

$$\epsilon_k = \frac{v^2}{2}$$

Since energy is conserved,

$$\epsilon = \epsilon_k + \epsilon_p,$$

and so the total specific orbital energy

$$\epsilon = \frac{v^2}{2} - \frac{GM}{r}$$

does not depend on the distance, r, from the center of the central body to the space vehicle in question. Therefore, the object can reach infinite r only if this quantity is nonnegative, which implies

$$v \geq \sqrt{\frac{2GM}{r}}.$$

The escape velocity from the Earth's surface is about 11 km/s, but that is insufficient to send the body an infinite distance because of the gravitational pull of the Sun. To escape the Solar System from a location at a distance from the Sun equal to the distance Sun–Earth, but not close to the Earth, requires around 42 km/s velocity, but there will be "part credit" for the Earth's orbital velocity for spacecraft launched from Earth, if their further acceleration (due to the propulsion system) carries them in the same direction as Earth travels in its orbit.

Formulae for Free Orbits

Orbits are conic sections, so the formula for the distance of a body for a given angle corresponds to the formula for that curve in polar coordinates, which is:

$$r = \frac{p}{1 + e\cos\theta}$$

$$\mu = G(m_1 + m_2)$$

$$p = h^2 / \mu$$

μ is called the gravitational parameter. m_1 and m_2 are the masses of objects 1 and 2, and h is the specific angular momentum of object 2 with respect to object 1. The parameter θ is known as the true anomaly, p is the semi-latus rectum, while e is the orbital eccentricity, all obtainable from the various forms of the six independent orbital elements.

Circular Orbits

All bounded orbits where the gravity of a central body dominates are elliptical in nature. A special case of this is the circular orbit, which is an ellipse of zero eccentricity. The formula for the velocity of a body in a circular orbit at distance r from the center of gravity of mass M can be derived as follows -

Centrifugal acceleration matches the acceleration due to gravity.

So, $v^2 / r = GM / r^2$

Therefore,

$$v = \sqrt{\frac{GM}{r}}$$

where G is the gravitational constant, equal to

$$6.673\,84 \times 10^{-11}\ \mathrm{m^3/(kg{\cdot}s^2)}$$

To properly use this formula, the units must be consistent; for example, M must be in kilograms, and r must be in meters. The answer will be in meters per second.

The quantity GM is often termed the standard gravitational parameter, which has a different value for every planet or moon in the Solar System.

Once the circular orbital velocity is known, the escape velocity is easily found by multiplying by the square root of 2:

$$v = \sqrt{2}\sqrt{\frac{GM}{r}} = \sqrt{\frac{2GM}{r}}.$$

To escape from gravity, the kinetic energy must at least match the negative potential energy. So, $1/2mv^2 = GMm / r$ and therefore,

$$v = \sqrt{\frac{2GM}{r}}.$$

Elliptical Orbits

If $0 < e < 1$, then the denominator of the equation of free orbits varies with the true anomaly θ, but remains positive, never becoming zero. Therefore, the relative position vector remains bounded, having its smallest magnitude at periapsis r_p, which is given by:

$$r_p = \frac{p}{1+e}$$

The maximum value r is reached when $\theta = 180^\circ$. This point is called the apoapsis, and its radial coordinate, denoted r_a, is

$$r_a = \frac{p}{1-e}$$

Let $2a$ be the distance measured along the apse line from periapsis P to apoapsis A, as illustrated in the equation below:

$$2a = r_p + r_a$$

Substituting the equations above, we get:

$$a = \frac{p}{1-e^2}$$

a is the semimajor axis of the ellipse. Solving for p, and substituting the result in the conic section curve formula above, we get:

$$r = \frac{a(1-e^2)}{1+e\cos\theta}$$

Orbital Period

Under standard assumptions the orbital period (T) of a body traveling along an elliptic orbit can be computed as:

$$T = 2\pi\sqrt{\frac{a^3}{\mu}}$$

where:

- μ is standard gravitational parameter,
- a is length of semi-major axis.

Conclusions:

- The orbital period is equal to that for a circular orbit with the orbit radius equal to the semi-major axis (a),
- For a given semi-major axis the orbital period does not depend on the eccentricity

Velocity

Under standard assumptions the orbital speed (v) of a body traveling along an elliptic orbit can be computed from the Vis-viva equation as:

$$v = \sqrt{\mu\left(\frac{2}{r} - \frac{1}{a}\right)}$$

where:

- μ is the standard gravitational parameter,
- r is the distance between the orbiting bodies.
- a is the length of the semi-major axis.

The velocity equation for a hyperbolic trajectory has either $+\frac{1}{a}$, or it is the same with the convention that in that case a is negative.

Energy

Under standard assumptions, specific orbital energy (ϵ) of elliptic orbit is negative and the orbital energy conservation equation (the Vis-viva equation) for this orbit can take the form:

$$\frac{v^2}{2} - \frac{\mu}{r} = -\frac{\mu}{2a} = \epsilon < 0$$

where:

- v is the speed of the orbiting body,
- r is the distance of the orbiting body from the center of mass of the central body,
- a is the semi-major axis,
- μ is the standard gravitational parameter.

Conclusions:

- For a given semi-major axis the specific orbital energy is independent of the eccentricity.

Using the virial theorem we find:

- the time-average of the specific potential energy is equal to 2ε
    - the time-average of r^{-1} is a^{-1}
- the time-average of the specific kinetic energy is equal to -ε

Parabolic Orbits

If the eccentricity equals 1, then the orbit equation becomes:

$$r = \frac{h^2}{\mu} \frac{1}{1 + \cos\theta}$$

where:

- r is the radial distance of the orbiting body from the mass center of the central body,
- h is specific angular momentum of the orbiting body,
- θ is the true anomaly of the orbiting body,
- μ is the standard gravitational parameter.

As the true anomaly θ approaches 180°, the denominator approaches zero, so that r tends towards infinity. Hence, the energy of the trajectory for which e=1 is zero, and is given by:

$$\epsilon = \frac{v^2}{2} - \frac{\mu}{r} = 0$$

where:

- v is the speed of the orbiting body.

In other words, the speed anywhere on a parabolic path is:

$$\sqrt{\quad}$$

Hyperbolic orbits

If $e > 1$, the orbit formula,

$$r = \frac{h^2}{\mu} \frac{1}{1 + e\cos\theta}$$

describes the geometry of the hyperbolic orbit. The system consists of two symmetric curves. The orbiting body occupies one of them; the other one is its empty mathematical image. Clearly, the denominator of the equation above goes to zero when $\cos\theta = -1/e$. we denote this value of true anomaly

$$\theta_\infty = \cos^{-1}\left(-\frac{1}{e}\right)$$

since the radial distance approaches infinity as the true anomaly approaches θ_∞ , known as the *true anomaly of the asymptote*. Observe that θ_∞ lies between 90° and 180°. From the trigonometric identity $\sin^2\theta + \cos^2\theta = 1$ it follows that:

$$\sin\theta_\infty = \frac{1}{e}\sqrt{e^2 - 1}$$

Energy

Under standard assumptions, specific orbital energy (ϵ) of a hyperbolic trajectory is greater than zero and the orbital energy conservation equation for this kind of trajectory takes form:

$$\epsilon = \frac{v^2}{2} - \frac{\mu}{r} = \frac{\mu}{-2a}$$

where:

- v is the orbital velocity of orbiting body,
- r is the radial distance of orbiting body from central body,
- a is the semi-major axis,
- μ is standard gravitational parameter.

Hyperbolic Excess Velocity

Under standard assumptions the body traveling along hyperbolic trajectory will attain in infinity an orbital velocity called hyperbolic excess velocity (v_∞) that can be computed as:

$$v_\infty = \sqrt{\frac{\mu}{-a}}$$

where:

- μ is standard gravitational parameter,
- a is the negative semi-major axis of orbit's hyperbola.

The hyperbolic excess velocity is related to the specific orbital energy or characteristic energy by

$$2\epsilon = C_3 = v_\infty^2$$

Calculating Trajectories

Kepler's Equation

One approach to calculating orbits (mainly used historically) is to use Kepler's equation:

$$M = E - \epsilon \cdot \sin E .$$

where M is the mean anomaly, E is the eccentric anomaly, and ϵ is the eccentricity.

With Kepler's formula, finding the time-of-flight to reach an angle (true anomaly) of θ from periapsis is broken into two steps:

- Compute the eccentric anomaly E from true anomaly θ
- Compute the time-of-flight t from the eccentric anomaly E

Finding the eccentric anomaly at a given time (the inverse problem) is more difficult. Kepler's equation is transcendental in E, meaning it cannot be solved for E algebraically. Kepler's equation can be solved for E analytically by inversion.

A solution of Kepler's equation, valid for all real values of ϵ is:

$$E = \begin{cases} \displaystyle\sum_{n=1}^{\infty} \frac{M^{\frac{n}{3}}}{n!} \lim_{\theta \to 0} \left(\frac{\mathrm{d}^{n-1}}{\mathrm{d}\theta^{n-1}} \left(\frac{\theta}{\sqrt[3]{\theta - \sin(\theta)}}^{n} \right) \right), & \epsilon = 1 \\ \displaystyle\sum_{n=1}^{\infty} \frac{M^{n}}{n!} \lim_{\theta \to 0} \left(\frac{\mathrm{d}^{n-1}}{\mathrm{d}\theta^{n-1}} \left(\frac{\theta}{\theta - \epsilon \cdot \sin(\theta)}^{n} \right) \right), & \epsilon \neq 1 \end{cases}$$

Evaluating this yields:

$$E = \begin{cases} x + \frac{1}{60}x^3 + \frac{1}{1400}x^5 + \frac{1}{25200}x^7 + \frac{43}{17248000}x^9 + \frac{1213}{7207200000}x^{11} + \frac{151439}{12713500800000}x^{13} \cdots \mid x = (6M)^{\frac{1}{3}}, & \epsilon = 1 \\ \frac{1}{1-\epsilon}M - \frac{\epsilon}{(1-\epsilon)^4}\frac{M^3}{3!} + \frac{(9\epsilon^2+\epsilon)}{(1-\epsilon)^7}\frac{M^5}{5!} - \frac{(225\epsilon^3+54\epsilon^2+\epsilon)}{(1-\epsilon)^{10}}\frac{M^7}{7!} + \frac{(11025\epsilon^4+4131\epsilon^3+243\epsilon^2+\epsilon)}{(1-\epsilon)^{13}}\frac{M^9}{9!}\cdots, & \epsilon \neq 1 \end{cases}$$

Alternatively, Kepler's Equation can be solved numerically. First one must guess a value of E and solve for time-of-flight; then adjust E as necessary to bring the computed time-of-flight closer to the desired value until the required precision is achieved. Usually, Newton's method is used to achieve relatively fast convergence.

The main difficulty with this approach is that it can take prohibitively long to converge for the extreme elliptical orbits. For near-parabolic orbits, eccentricity ϵ is nearly 1, and plugging $e=1$ into the formula for mean anomaly, $E - \sin E$, we find ourselves subtracting two nearly-equal values, and accuracy suffers. For near-circular orbits, it is hard to find the periapsis in the first place (and truly circular orbits have no periapsis at all). Furthermore, the equation was derived on the assumption of an elliptical orbit, and so it does not hold for parabolic or hyperbolic orbits. These difficulties are what led to the development of the universal variable formulation, described below.

Conic Orbits

For simple procedures, such as computing the delta-v for coplanar transfer ellipses, traditional approaches are fairly effective. Others, such as time-of-flight are far more complicated, especially for near-circular and hyperbolic orbits.

The Patched Conic Approximation

The Hohmann transfer orbit alone is a poor approximation for interplanetary trajectories because it neglects the planets' own gravity. Planetary gravity dominates the behaviour of the spacecraft in the vicinity of a planet and in most cases Hohmann severely overestimates delta-v, and produces highly inaccurate prescriptions for burn timings.

A relatively simple way to get a first-order approximation of delta-v is based on the 'Patched Conic Approximation' technique. One must choose the one dominant gravitating body in each region of space through which the trajectory will pass, and to model only that body's effects in that region. For instance, on a trajectory from the Earth to Mars, one would begin by considering only the Earth's gravity until the trajectory reaches a distance where the Earth's gravity no longer dominates that of the Sun. The spacecraft would be given escape velocity to send it on its way to interplanetary space. Next, one would consider only the Sun's gravity until the trajectory reaches the neighbourhood of Mars. During this stage, the transfer orbit model is appropriate. Finally, only Mars's gravity is considered during the final portion of the trajectory where Mars's gravity dominates the spacecraft's behaviour. The spacecraft would approach Mars on a hyperbolic orbit, and a final retrograde burn would slow the spacecraft enough to be captured by Mars.

The size of the "neighbourhoods" (or spheres of influence) vary with radius r_{SOI}:

$$r_{SOI} = a_p \left(\frac{m_p}{m_s} \right)^{2/5}$$

where a_p is the semimajor axis of the planet's orbit relative to the Sun; m_p and m_s are the masses of the planet and Sun, respectively.

This simplification is sufficient to compute rough estimates of fuel requirements, and rough time-of-flight estimates, but it is not generally accurate enough to guide a spacecraft to its destination. For that, numerical methods are required.

The Universal Variable Formulation

To address computational shortcomings of traditional approaches for solving the 2-body problem, the universal variable formulation was developed. It works equally well for the circular, elliptical, parabolic, and hyperbolic cases, the differential equations converging well when integrated for any orbit. It also generalizes well to problems incorporating perturbation theory.

Perturbations

The universal variable formulation works well with the variation of parameters technique, except now, instead of the six Keplerian orbital elements, we use a different set of orbital elements: namely, the satellite's initial position and velocity vectors x_0 and v_0 at a given epoch $t = 0$. In a two-body simulation, these elements are sufficient to compute the satellite's position and velocity at any time in the future, using the universal variable formulation. Conversely, at any moment in the satellite's orbit, we can measure its position and velocity, and then use the universal variable approach to determine what its initial position and velocity *would have been* at the epoch. In perfect two-body motion, these orbital elements would be invariant (just like the Keplerian elements would be).

However, perturbations cause the orbital elements to change over time. Hence, we write the position element as $x_0(t)$ and the velocity element as $v_0(t)$, indicating that they vary with time. The technique to compute the effect of perturbations becomes one of finding expressions, either exact or approximate, for the functions $x_0(t)$ and $v_0(t)$.

The following are some effects which make real orbits differ from the simple models based on a spherical earth. Most of them can be handled on short timescales (perhaps less than a few thousand orbits) by perturbation theory because they are small relative to the corresponding two-body effects.

- Equatorial bulges cause precession of the node and the perigee
- Tesseral harmonics of the gravity field introduce additional perturbations
- Lunar and solar gravity perturbations alter the orbits
- Atmospheric drag reduces the semi-major axis unless make-up thrust is used

Over very long timescales (perhaps millions of orbits), even small perturbations can dominate, and the behaviour can become chaotic. On the other hand, the various perturbations can be orchestrated by clever astrodynamicists to assist with orbit maintenance tasks, such as station-keeping, ground track maintenance or adjustment, or phasing of perigee to cover selected targets at low altitude.

Orbital Maneuver

In spaceflight, an orbital maneuver is the use of propulsion systems to change the orbit of a spacecraft. For spacecraft far from Earth for example those in orbits around the Sun an orbital maneuver is called a *deep-space maneuver (DSM)*.

Orbital Transfer

Transfer orbits are usually elliptical orbits that allow spacecraft to move from one (usually substantially circular) orbit to another. Usually they require a burn at the start, a burn at the end, and sometimes one or more burns in the middle.

- The Hohmann transfer orbit requires a minimal delta-v.
- A bi-elliptic transfer can require less energy than the Hohmann transfer, if the ratio of orbits is 11.94 or greater, but comes at the cost of increased trip time over the Hohmann transfer.
- Faster transfers may use any orbit that intersects both the original and destination orbits, at the cost of higher delta-v.
- Using low thrust engines (such as electrical propulsion), if the initial orbit is supersynchronous to the final desired circular orbit then the optimal transfer orbit is achieved by thrusting continuously in the direction of the velocity at apogee. This method however takes much longer due to the low thrust.

For the case of orbital transfer between non-coplanar orbits, the change-of-plane thrust must be made at the point where the orbital planes intersect (the "node").

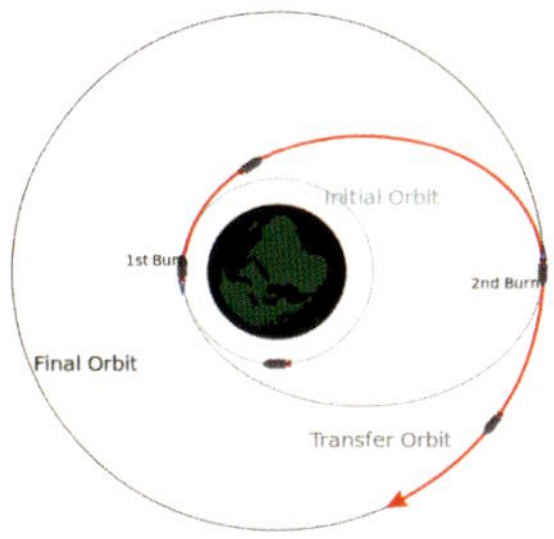

A Hohmann transfer from a low circular orbit to a higher circular orbit

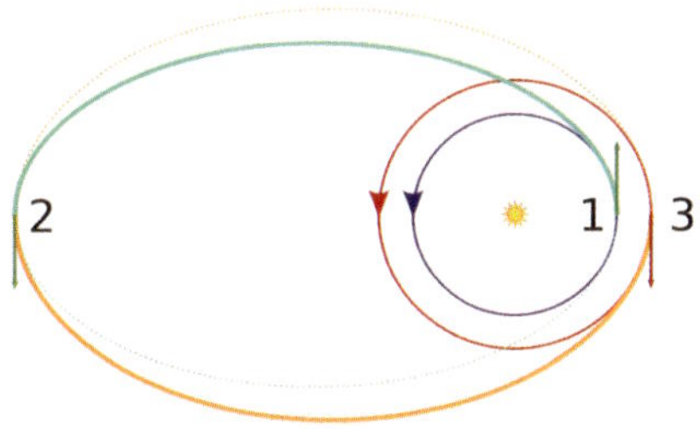

A bi-elliptic transfer from a low circular starting orbit (dark blue), to a higher circular orbit (red)

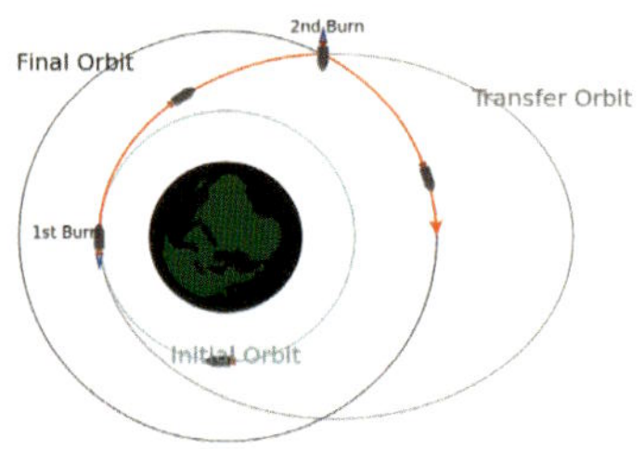

Generic two-impulse elliptical transfer between two circular orbits

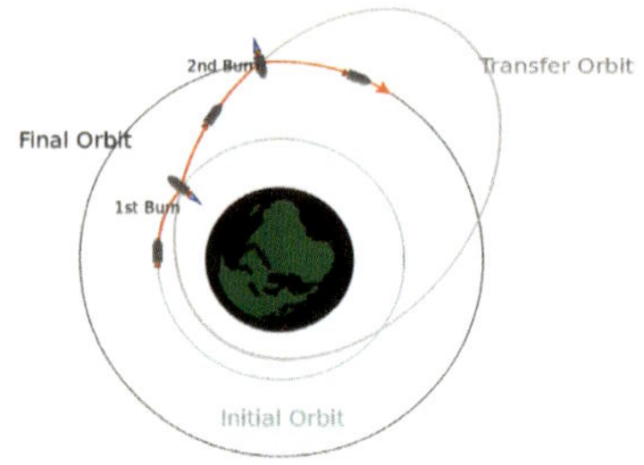

A general transfer from a low circular orbit to a higher circular orbit

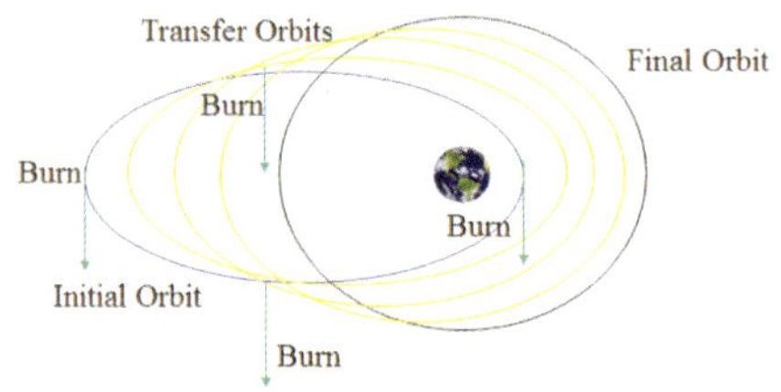

An optimal sequence for transferring a satellite from a supersynchronous to a geosynchronous orbit using electric propulsion

Gravity Assist and the Oberth Effect

In a gravity assist, a spacecraft swings by a planet and leaves in a different direction, at a different speed. This is useful to speed or slow a spacecraft instead of carrying more fuel.

This maneuver can be approximated by an elastic collision at large distances, though the flyby does not involve any physical contact. Due to Newton's Third Law (equal and opposite reaction), any momentum gained by a spacecraft must be lost by the planet, or vice versa. However, because the planet is much, much more massive than the spacecraft, the effect on the planet's orbit is negligible.

The Oberth effect can be employed, particularly during a gravity assist operation. This effect is that use of a propulsion system works better at high speeds, and hence course changes are best done when close to a gravitating body; this can multiply the effective delta-v.

Interplanetary Transport Network and Fuzzy Orbits

It is now possible to use computers to search for routes using the nonlinearities in the gravity of the planets and moons of the Solar System. For example, it is possible to plot an orbit from high earth orbit to Mars, passing close to one of the Earth's Trojan points. Collectively referred to as the Interplanetary Transport Network, these highly perturbative, even chaotic, orbital trajectories in principle need no fuel beyond that needed to reach the Lagrange point (in practice keeping to the trajectory requires some course corrections). The biggest problem with them is they can be exceedingly slow, taking many years. In addition launch windows can be very far apart.

They have, however, been employed on projects such as Genesis. This spacecraft visited the Earth-Sun L_1 point and returned using very little propellant.

Kepler's Laws of Planetary Motion

Kepler's laws of planetary motion, in astronomy and classical physics are the laws describing the motions of the planets in the solar system. They were derived by the German astronomer Johannes Kepler, whose analysis of the observations of the 16th-century Danish astronomer Tycho Brahe enabled him to announce his first two laws in the year 1609 and a third law nearly a decade later, in 1618. Kepler himself never numbered these laws or specially distinguished them from his other discoveries.

Kepler's three laws of planetary motion can be stated as follows: (1) All planets move about the Sun in elliptical orbits, having the Sun as one of the foci. (2) A radius vector joining any planet to the Sun sweeps out equal areas in equal lengths of time. (3) The squares of the sidereal periods (of revolution) of the planets are directly proportional to the cubes of their mean distances from the Sun. Knowledge of these laws, especially the second (the law of areas), proved crucial to Sir Isaac Newton in 1684–85, when he formulated his famous law of gravitation between Earth and the Moon and between the Sun and the planets, postulated by him to have validity for all objects anywhere in the universe. Newton showed that the motion of bodies subject to central gravitational force need not always follow the elliptical orbits specified by the first law of Kepler but can take paths defined by other, open conic curves; the motion can be in parabolic or hyperbolic orbits, depending on the total energy of the body. Thus, an object of sufficient energy e.g., a comet can enter the solar system and leave again without returning. From Kepler's second law, it may be observed further that the angular momentum of any planet about an axis through the Sun and perpendicular to the orbital plane is also unchanging.

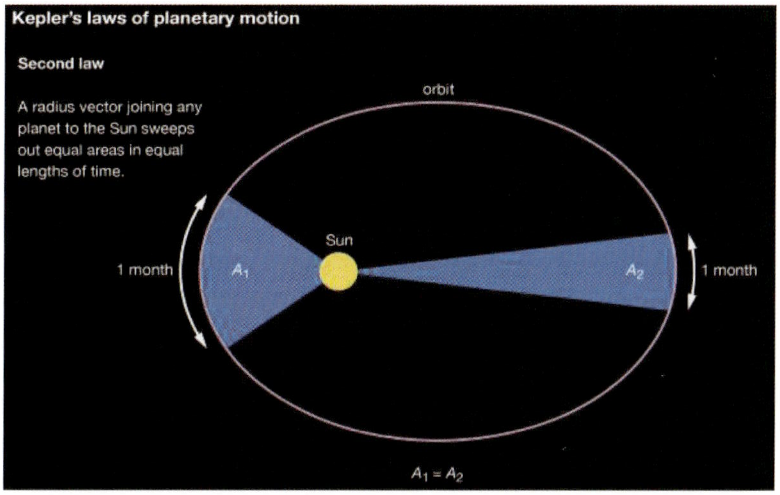

Kepler's second law of planetary motion: A radius vector joining any planet to the Sun sweeps out equal areas in equal lengths of time

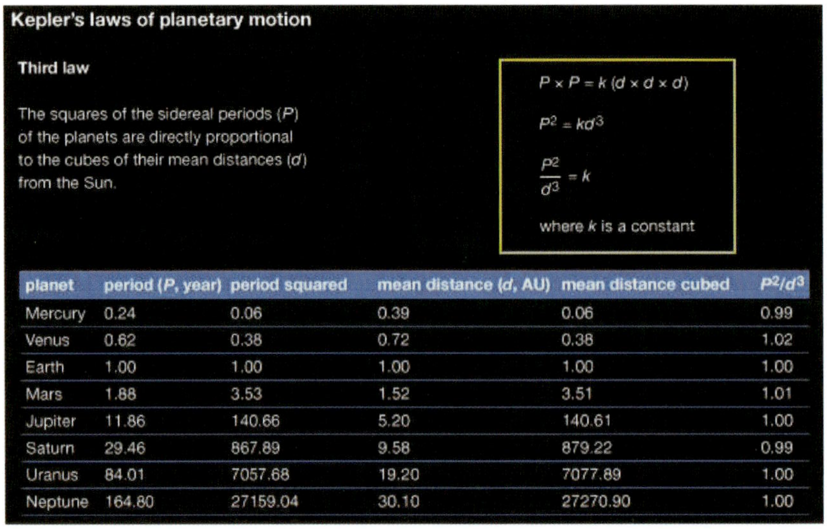

Kepler's laws of planetary motion

Third law

The squares of the sidereal periods (P) of the planets are directly proportional to the cubes of their mean distances (d) from the Sun.

$P \times P = k\,(d \times d \times d)$

$P^2 = kd^3$

$\frac{P^2}{d^3} = k$

where k is a constant

planet	period (P, year)	period squared	mean distance (d, AU)	mean distance cubed	P^2/d^3
Mercury	0.24	0.06	0.39	0.06	0.99
Venus	0.62	0.38	0.72	0.38	1.02
Earth	1.00	1.00	1.00	1.00	1.00
Mars	1.88	3.53	1.52	3.51	1.01
Jupiter	11.86	140.66	5.20	140.61	1.00
Saturn	29.46	867.89	9.58	879.22	0.99
Uranus	84.01	7057.68	19.20	7077.89	1.00
Neptune	164.80	27159.04	30.10	27270.90	1.00

Kepler's third lawKepler's third law of planetary motion. The squares of the sidereal periods (P) of the planets are directly proportional to the cubes of their mean distances (d) from the Sun

Formulary

The mathematical model of the kinematics of a planet subject to the laws allows a large range of further calculations.

First law of Kepler

The orbit of every planet is an ellipse with the Sun at one of the two foci.

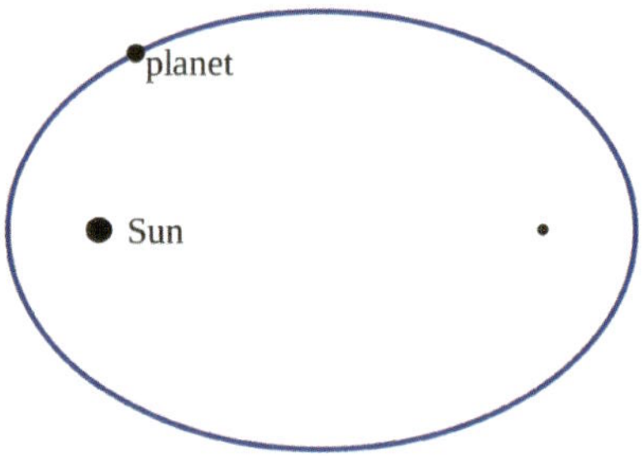

Figure: Kepler's first law placing the Sun at the focus of an elliptical orbit

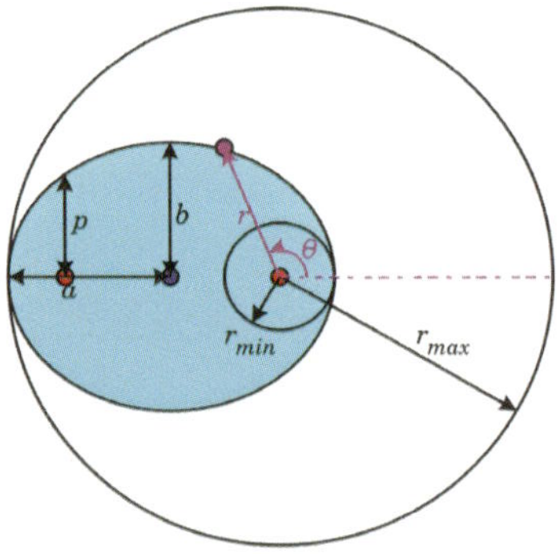

Figure: Heliocentric coordinate system (r, θ) for ellipse. Also shown are: semi-major axis a, semi-minor axis b and semi-latus rectum p; center of ellipse and its two foci marked by large dots. For $\theta = 0°$, $r = r_{min}$ and for $\theta = 180°$, $r = r_{max}$

Mathematically, an ellipse can be represented by the formula:

$$r = \frac{p}{1+\varepsilon \cos\theta}$$

where p is the semi-latus rectum, ε is the eccentricity of the ellipse, r is the distance from the Sun to the planet, and θ is the angle to the planet's current position from its closest approach, as seen from the Sun. So (r, θ) are polar coordinates.

For an eclipse $0 < \varepsilon < 1$; in the limiting case $\varepsilon = 0$, the orbit is a circle with the sun at the centre (i.e. where there is zero eccentricity)

At $\theta = 0°$, perihelion, the distance is minimum

$$r_{min} = \frac{p}{1+\varepsilon}$$

At $\theta = 90°$ and at $\theta = 270°$ the distance is equal to p.

At $\theta = 180°$, aphelion, the distance is maximum (by definition, aphelion is – invariably – perihelion plus 180°)

$$r_{max} = \frac{p}{1-\varepsilon}$$

The semi-major axis a is the arithmetic mean between r_{min} and r_{max}:

$$r_{max} - a = a - r_{min}$$

$$a = \frac{p}{1-\varepsilon^2}$$

The semi-minor axis b is the geometric mean between r_{min} and r_{max}:

$$\frac{r_{max}}{b} = \frac{b}{r_{min}}$$

$$b = \frac{p}{\sqrt{1-\varepsilon^2}}$$

The semi-latus rectum p is the harmonic mean between r_{min} and r_{max}:

$$\frac{1}{r_{min}} - \frac{1}{p} = \frac{1}{p} - \frac{1}{r_{max}}$$

$$pa = r_{max} r_{min} = b^2$$

The eccentricity ε is the coefficient of variation between r_{min} and r_{max}:

$$\varepsilon = \frac{r_{max} - r_{min}}{r_{max} + r_{min}}.$$

The area of the ellipse is

$$A = \pi ab.$$

The special case of a circle is $\varepsilon = 0$, resulting in $r = p = r_{min} = r_{max} = a = b$ and $A = \pi r^2$.

Second Law of Kepler

A line joining a planet and the Sun sweeps out equal areas during equal intervals of time.

The orbital radius and angular velocity of the planet in the elliptical orbit will vary. This is shown in the animation: the planet travels faster when closer to the sun, then slower when farther from the sun. Kepler's second law states that the blue sector has constant area.

In a small time dt the planet sweeps out a small triangle having base line r and height $rd\theta$ and area $dA = \frac{1}{2} \cdot r \cdot rd\theta$ and so the constant areal velocity is $\frac{dA}{dt} = \frac{1}{2} r^2 \frac{d\theta}{dt}$.
The area enclosed by the elliptical orbit is πab. So the period P satisfies

$$P \cdot \tfrac{1}{2} r^2 \frac{d\theta}{dt} = \pi ab$$

and the mean motion of the planet around the Sun

$$n = 2\pi / P$$

satisfies

$$r^2 d\theta = abn\ dt.$$

Third law of Kepler

The square of the orbital period of a planet is directly proportional to the cube of the semi-major axis of its orbit.

This captures the relationship between the distance of planets from the Sun, and their orbital periods.

Kepler enunciated in 1619 this third law in a laborious attempt to determine what he viewed as the "music of the spheres" according to precise laws, and express it in terms of musical notation. So it was known as the *harmonic law*.

Using Newton's Law of gravitation, this relation can be found in the case of a circular orbit by setting the centripetal force equal to the gravitational force:

$$mr_{\omega}^{\ 2} = G\frac{mM}{r^2}$$

Then, expressing the angular velocity in terms of the orbital period and then rearranging, we find Kepler's Third Law:

$$mr\left(\frac{2\pi}{T}\right)^2 = G\frac{mM}{r^2} \rightarrow T^2 = \left(\frac{4\pi^2}{GM}\right)r^3 \rightarrow T^2 \propto r^3$$

A more detailed derivation can be done with general elliptical orbits, instead of circles, as well as orbiting the center of mass, instead of just the large mass. This results in replacing a circular radius, r, with the elliptical semi-major axis, a, as well as replacing the large mass M with $M+m$. However, with planet masses being so much smaller than the sun, this correction is often ignored. The full corresponding formula is:

$$\frac{a^3}{T^2} = \frac{G(M+m)}{4\pi^2} \approx \frac{GM}{4\pi^2} = 7.495 \cdot 10^{-6}\left[\frac{\text{AU}^3}{\text{days}^2}\right] = \text{constant}$$

where M is the mass of the sun, m is the mass of the planet, and G is the gravitational constant, T is the orbital period and a is the elliptical semi-major axis.

The following table shows the data used by Kepler to empirically derive his law:

Data used by Kepler (1618)			
Planet	**Mean distance to sun [AU]**	**Period [days]**	$R^3 / T^2 \cdot 10^6 [\text{AU}^3 / \text{day}^2]$
Mercury	0.389	87.77	7.64
Venus	0.724	224.70	7.52
Earth	1	365.25	7.50
Mars	1.524	686.95	7.50
Jupiter	5.2	4332.62	7.49
Saturn	9.510	10759.2	7.43

Upon finding this pattern Kepler wrote:

"I first believed I was dreaming. But it is absolutely certain and exact that the ratio which exists between the period times of any two planets is precisely the ratio of the 3/2th power of the mean distance."

translated from "Harmonies of the World" by Kepler (1619)

For comparison, here are modern estimates:

Modern data (Wolfram Alpha Knowledgebase 2018)			
Planet	**Semi-major axis [AU]**	**Period [days]**	$R^3 / T^2 \cdot 10^6 [\text{AU}^3 / \text{day}^2]$.
Mercury	0.38710	87.9693	7.496
Venus	0.72333	224.7008	7.496
Earth	1	365.2564	7.496
Mars	1.52366	686.9796	7.495

Jupiter	5.20336	4332.8201	7.504
Saturn	9.53707	10775.599	7.498
Uranus	19.1913	30687.153	7.506
Neptune	30.0690	60190.03	7.504

Planetary Acceleration

Isaac Newton computed in his *Philosophiæ Naturalis Principia Mathematica* the acceleration of a planet moving according to Kepler's first and second law.

1. The *direction* of the acceleration is towards the Sun.
2. The *magnitude* of the acceleration is inversely proportional to the square of the planet's distance from the Sun (the *inverse square law*).

This implies that the Sun may be the physical cause of the acceleration of planets. However, Newton states in his Principia that he considers forces from a mathematical point of view, not a physical, thereby taking an instrumentalist view. Moreover, he does not assign a cause to gravity.

Newton defined the force acting on a planet to be the product of its mass and the acceleration. So:

1. Every planet is attracted towards the Sun.
2. The force acting on a planet is directly proportional to the mass of the planet and is inversely proportional to the square of its distance from the Sun.

The Sun plays an unsymmetrical part, which is unjustified. So he assumed, in Newton's law of universal gravitation:

1. All bodies in the solar system attract one another.
2. The force between two bodies is in direct proportion to the product of their masses and in inverse proportion to the square of the distance between them.

As the planets have small masses compared to that of the Sun, the orbits conform approximately to Kepler's laws. Newton's model improves upon Kepler's model, and fits actual observations more accurately.

Below comes the detailed calculation of the acceleration of a planet moving according to Kepler's first and second laws.

Acceleration Vector

From the heliocentric point of view consider the vector to the planet $\mathrm{r} = r\hat{\mathrm{r}}$ where r is the distance to the planet and $\hat{\mathrm{r}}$ is a unit vector pointing towards the planet.

$$\frac{d\hat{\mathrm{r}}}{dt} = \dot{\hat{\mathrm{r}}} = \dot{\theta}\,\hat{\theta}, \qquad \frac{d\hat{\theta}}{dt} = \dot{\hat{\theta}} = -\dot{\theta}\,\hat{\mathrm{r}}$$

where $\hat{\theta}$ is the unit vector whose direction is 90 degrees counterclockwise of $\hat{\mathrm{r}}$, and θ is the polar angle, and where a dot on top of the variable signifies differentiation with respect to time.

Differentiate the position vector twice to obtain the velocity vector and the acceleration vector:

$$\dot{\mathrm{r}} = \dot{r}\hat{\mathrm{r}} + r\dot{\hat{\mathrm{r}}} = \dot{r}\hat{\mathrm{r}} + r\dot{\theta}\hat{\theta},$$

$$\ddot{\mathrm{r}} = (\ddot{r}\hat{\mathrm{r}} + \dot{r}\dot{\hat{\mathrm{r}}}) + (\dot{r}\dot{\theta}\,\hat{\theta} + r\ddot{\theta}\,\hat{\theta} + r\dot{\theta}\dot{\hat{\theta}}) = (\ddot{r} - r\dot{\theta}^2)\hat{\mathrm{r}} + (r\ddot{\theta} + 2\dot{r}\dot{\theta})\hat{\theta}.$$

So

$$\ddot{\mathrm{r}} = a_r\hat{\mathrm{r}} + a_\theta\hat{\theta}$$

where the radial acceleration is

$$a_r = \ddot{r} - r\dot{\theta}^2$$

and the transversal acceleration is

$$a_\theta = r\ddot{\theta} + 2\dot{r}\dot{\theta}.$$

Inverse Square Law

Kepler's second law says that

$$r^2\dot{\theta} = nab$$

is constant.

The transversal acceleration a_θ is zero:

$$\frac{d(r^2\dot{\theta})}{dt} = r(2\dot{r}\dot{\theta} + r\ddot{\theta}) = ra_\theta = 0.$$

So the acceleration of a planet obeying Kepler's second law is directed towards the sun.

The radial acceleration a_r is

$$a_r = \ddot{r} - r\dot{\theta}^2 = \ddot{r} - r\left(\frac{nab}{r^2}\right)^2 = \ddot{r} - \frac{n^2a^2b^2}{r^3}.$$

Kepler's first law states that the orbit is described by the equation:

$$\frac{p}{r} = 1 + \varepsilon\cos\theta$$

Differentiating with respect to time

$$-\frac{p\dot{r}}{r^2} = -\varepsilon\sin\theta\dot{\theta}$$

or

$$p\dot{r} = nab\varepsilon\sin\theta.$$

Differentiating once more

$$p\ddot{r} = nab\varepsilon\cos\theta\dot{\theta} = nab\varepsilon\cos\theta\frac{nab}{r^2} = \frac{n^2a^2b^2}{r^2}\varepsilon\cos\theta$$

The radial acceleration a_r satisfies

$$pa_r = \frac{n^2a^2b^2}{r^2}\varepsilon\cos\theta - p\frac{n^2a^2b^2}{r^3} = \frac{n^2a^2b^2}{r^2}\left(\varepsilon\cos\theta - \frac{p}{r}\right).$$

Substituting the equation of the ellipse gives

$$pa_r = \frac{n^2a^2b^2}{r^2}\left(\frac{p}{r} - 1 - \frac{p}{r}\right) = -\frac{n^2a^2}{r^2}b^2.$$

The relation $b^2 = pa$ gives the simple final result

$$a_r = -\frac{n^2a^3}{r^2}.$$

This means that the acceleration vector $\ddot{\mathbf{r}}$ of any planet obeying Kepler's first and second law satisfies the inverse square law

$$\ddot{\mathbf{r}} = -\frac{\alpha}{r^2}\hat{\mathbf{r}}$$

where

$$\alpha = n^2a^3$$

is a constant, and $\hat{\mathbf{r}}$ is the unit vector pointing from the Sun towards the planet, and r is the distance between the planet and the Sun.

According to Kepler's third law, α has the same value for all the planets. So the inverse square law for planetary accelerations applies throughout the entire solar system.

The inverse square law is a differential equation. The solutions to this differential equation include the Keplerian motions, as shown, but they also include motions where the orbit is a hyperbola or parabola or a straight line.

Newton's Law of Gravitation

By Newton's second law, the gravitational force that acts on the planet is:

$$\mathbf{F} = m_{\text{planet}}\ddot{\mathbf{r}} = -m_{\text{planet}}\alpha r^{-2}\hat{\mathbf{r}}$$

where m_{planet} is the mass of the planet and α has the same value for all planets in the solar system. According to Newton's third Law, the Sun is attracted to the planet by a force of the same magnitude.

Since the force is proportional to the mass of the planet, under the symmetric consideration, it should also be proportional to the mass of the Sun, m_{Sun}. So

$$\alpha = Gm_{\text{Sun}}$$

where G is the gravitational constant.

The acceleration of solar system body number i is, according to Newton's laws:

$$\ddot{\mathrm{r}}_{\mathrm{i}} = G\sum_{j\neq i} m_j r_{ij}^{-2}\,\hat{\mathrm{r}}_{ij}$$

where m_j is the mass of body j, r_{ij} is the distance between body i and body j, $\hat{\mathrm{r}}_{ij}$ is the unit vector from body i towards body j, and the vector summation is over all bodies in the world, besides i itself.

In the special case where there are only two bodies in the world, Earth and Sun, the acceleration becomes

$$\ddot{\mathrm{r}}_{\text{Earth}} = Gm_{\text{Sun}} r_{\text{Earth,Sun}}^{-2}\,\hat{\mathrm{r}}_{\text{Earth,Sun}}$$

which is the acceleration of the Kepler motion. So this Earth moves around the Sun according to Kepler's laws.

If the two bodies in the world are Moon and Earth the acceleration of the Moon becomes

$$\ddot{\mathrm{r}}_{\text{Moon}} = Gm_{\text{Earth}} r_{\text{Moon,Earth}}^{-2}\,\hat{\mathrm{r}}_{\text{Moon,Earth}}$$

So in this approximation the Moon moves around the Earth according to Kepler's laws.

In the three-body case the accelerations are

$$\ddot{\mathrm{r}}_{\text{Sun}} = Gm_{\text{Earth}} r_{\text{Sun,Earth}}^{-2}\,\hat{\mathrm{r}}_{\text{Sun,Earth}} + Gm_{\text{Moon}} r_{\text{Sun,Moon}}^{-2}\,\hat{\mathrm{r}}_{\text{Sun,Moon}}$$

$$\ddot{\mathrm{r}}_{\text{Earth}} = Gm_{\text{Sun}} r_{\text{Earth,Sun}}^{-2}\,\hat{\mathrm{r}}_{\text{Earth,Sun}} + Gm_{\text{Moon}} r_{\text{Earth,Moon}}^{-2}\,\hat{\mathrm{r}}_{\text{Earth,Moon}}$$

$$\ddot{\mathrm{r}}_{\text{Moon}} = Gm_{\text{Sun}} r_{\text{Moon,Sun}}^{-2}\,\hat{\mathrm{r}}_{\text{Moon,Sun}} + Gm_{\text{Earth}} r_{\text{Moon,Earth}}^{-2}\,\hat{\mathrm{r}}_{\text{Moon,Earth}}$$

These accelerations are not those of Kepler orbits, and the three-body problem is complicated. But Keplerian approximation is the basis for perturbation calculations.

Position as a Function of Time

Kepler used his two first laws to compute the position of a planet as a function of time. His method involves the solution of a transcendental equation called Kepler's equation.

The procedure for calculating the heliocentric polar coordinates (r,θ) of a planet as a function of the time t since perihelion, is the following four steps:

1. Compute the mean anomaly M = nt where n is the mean motion.

$n \cdot P = 2\pi$ radians where P is the period.

2. Compute the eccentric anomaly E by solving Kepler's equation:

$$M = E - \varepsilon \sin E$$

3. Compute the true anomaly θ by the equation:

$$(1-\varepsilon)\tan^2\frac{\theta}{2} = (1+\varepsilon)\tan^2\frac{E}{2}$$

4. Compute the heliocentric distance

$$r = a(1-\varepsilon\cos E).$$

The Cartesian velocity vector can along be trivially calculated as $v = \frac{\sqrt{\mu a}}{r}\left\langle -\sin E, \sqrt{1-\varepsilon^2}\sin E \right\rangle$.

The important special case of circular orbit, $\varepsilon = 0$, gives $\theta = E = M$. Because the uniform circular motion was considered to be *normal*, a deviation from this motion was considered an anomaly.

The proof of this procedure is shown below.

Mean Anomaly, M

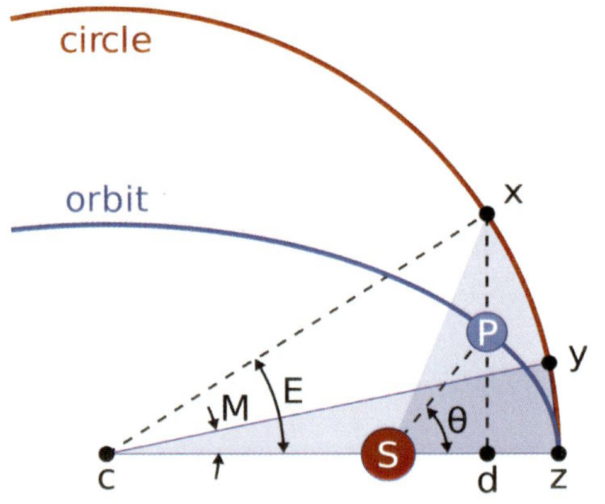

Figure: Geometric construction for Kepler's calculation of θ. The Sun (located at the focus) is labeled S and the planet P. The auxiliary circle is an aid to calculation. Line xd is perpendicular to the base and through the planet P. The shaded sectors are arranged to have equal areas by positioning of point y

The Keplerian problem assumes an elliptical orbit and the four points:

s the Sun (at one focus of ellipse);

z the perihelion

c the center of the ellipse

p the planet

and

$a = |cz|$, distance between center and perihelion, the semimajor axis,

$\varepsilon = \frac{|cs|}{a}$, the eccentricity,

$b = a\sqrt{1-\varepsilon^2}$, the semiminor axis,

$r = |sp|$, the distance between Sun and planet.

$\theta = \angle zsp$, the direction to the planet as seen from the Sun, the true anomaly.

The problem is to compute the polar coordinates (r,θ) of the planet from the time since perihelion, t.

It is solved in steps. Kepler considered the circle with the major axis as a diameter, and

x, the projection of the planet to the auxiliary circle

y, the point on the circle such that the sector areas $|zcy|$ and $|zsx|$ are equal,

$M = \angle zcy$, the mean anomaly.

The sector areas are related by $|zsp| = \frac{b}{a} \cdot |zsx|$.

The circular sector area $|zcy| = \frac{a^2 M}{2}$.

The area swept since perihelion,

$$|zsp| = \frac{b}{a} \cdot |zsx| = \frac{b}{a} \cdot |zcy| = \frac{b}{a} \cdot \frac{a^2 M}{2} = \frac{abM}{2},$$

is by Kepler's second law proportional to time since perihelion. So the mean anomaly, M, is proportional to time since perihelion, t.

$$M = nt,$$

where n is the mean motion.

Eccentric Anomaly, E

When the mean anomaly M is computed, the goal is to compute the true anomaly θ. The function $\theta = f(M)$ is, however, not elementary.Kepler's solution is to use

$E = \angle zcx$, x as seen from the centre, the eccentric anomaly

as an intermediate variable, and first compute E as a function of M by solving Kepler's equation below, and then compute the true anomaly θ from the eccentric anomaly E. Here are the details.

$$|zcy| = |zsx| = |zcx| - |scx|$$

$$\frac{a^2 M}{2} = \frac{a^2 E}{2} - \frac{a\varepsilon \cdot a \sin E}{2}$$

Division by $a^2/2$ gives Kepler's equation

$$M = E - \varepsilon \sin E.$$

This equation gives M as a function of E. Determining E for a given M is the inverse problem. Iterative numerical algorithms are commonly used.

Having computed the eccentric anomaly E, the next step is to calculate the true anomaly θ.

True anomaly, θ

Note from the figure that

$$\overrightarrow{cd} = \overrightarrow{cs} + \overrightarrow{sd}$$

so that

$$a \cos E = a\varepsilon + r \cos\theta.$$

Dividing by a and inserting from Kepler's first law

$$\frac{r}{a} = \frac{1-\varepsilon^2}{1+\varepsilon\cos\theta}$$

to get

$$\cos E = \varepsilon + \frac{1-\varepsilon^2}{1+\varepsilon\cos\theta}\cos\theta = \frac{\varepsilon(1+\varepsilon\cos\theta)+(1-\varepsilon^2)\cos\theta}{1+\varepsilon\cos\theta} = \frac{\varepsilon+\cos\theta}{1+\varepsilon\cos\theta}.$$

The result is a usable relationship between the eccentric anomaly E and the true anomaly θ.

A computationally more convenient form follows by substituting into the trigonometric identity:

$$\tan^2\frac{x}{2} = \frac{1-\cos x}{1+\cos x}.$$

Get

$$\tan^2\frac{E}{2} = \frac{1-\cos E}{1+\cos E} = \frac{1-\dfrac{\varepsilon+\cos\theta}{1+\varepsilon\cos\theta}}{1+\dfrac{\varepsilon+\cos\theta}{1+\varepsilon\cos\theta}}$$
$$= \frac{(1+\varepsilon\cos\theta)-(\varepsilon+\cos\theta)}{(1+\varepsilon\cos\theta)+(\varepsilon+\cos\theta)} = \frac{1-\varepsilon}{1+\varepsilon}\cdot\frac{1-\cos\theta}{1+\cos\theta} = \frac{1-\varepsilon}{1+\varepsilon}\tan^2\frac{\theta}{2}.$$

Multiplying by $1 + \varepsilon$ gives the result

$$(1-\varepsilon)\tan^2\frac{\theta}{2} = (1+\varepsilon)\tan^2\frac{E}{2}$$

This is the third step in the connection between time and position in the orbit.

Distance, r

The fourth step is to compute the heliocentric distance r from the true anomaly θ by Kepler's first law:

$$r(1+\varepsilon\cos\theta)=a(1-\varepsilon^2)$$

Using the relation above between θ and E the final equation for the distance r is:

$$r=a(1-\varepsilon\cos E).$$

Orbital Velocity

Orbital Velocity of a Earth satellite is the minimum velocity it has to maintain to continue its circular motion in its orbit. We know for a circular motion to continue, a force called centripetal force is required to work upon the object in circular motion. And this force is a real and acts towards the centre of the circle. As this is a real force it's to be supplied by a real mechanism or system. A satellite moves in a circular track and the centripetal force acting on it is supplied by the gravitational force acting between the earth and the satellite. Orbital Velocity formula:With the help of this concept or information we can easily go for the derivation of Orbital Velocity.

orbital velocity of satellite

Derivation of Orbital Velocity

Deriving the expression or formula 1

The Gravitational Force between the earth and the satellite = $F_g = (G.M.m)/r^2$

The centripetal force acting on the satellite = $F_c = mV^2/r$

Here, M is the mass of earth and m is the mass of the satellite which is having a uniform circular motion in a circular track of radius r around the earth. V is the linear velocity of the satellite at a point on its circular track.

Now this r is the sum of the radius of the earth(R) and the height(h) of the satellite from the surface of the earth.

$r = R + h$

Now equating, equation $F_g = (G.M.m)/r^2$ and $F_c = mV^2/r$ we get,

$$F_g = F_c$$

$$=> (G.M.m)/r^2 = mV^2/r$$

$$V = [(GM)/r]^{1/2}$$

This is the first equation or expression of Orbital Velocity of a satellite. Here $r = R + h$

Another Expression or Formula of Orbital Velocity of a Satellite

Deriving the expression or formula 2

For a mass of m on earth's surface, the following is true:

$mg = (GMm)/R^2$

Note, on earth surface h=0 and r = R. And gravitational force on a mass is equal to its weight on the surface.

From above equation we get this equation, $GM = g. R^2$

Substituting this expression of GM in equ. $V = \left(\frac{gm}{r}\right)^{1/2}$ (Orbital velocity), we get,

$$V = [(gR^2)/r]^{1/2}$$

$$V = R . (g/r)1/2$$

This is the second expression of Orbital Velocity of a satellite. Here, as said earlier, $r = R + h$

Next we will derive the 3rd equation and that is for a nearby orbit, i.e. for an orbit which has negligible height above the earth's surface.

Nearby Orbit and Orbital Velocity – when Height above Surface is Less or Reduced

Let's consider an orbit which is pretty close to the earth. (formula 3)

Now if the height of the satellite (h) from the surface of the earth is negligible with respect to the Radius of the earth, then we can write $r=R+h = R$ (as h is negligible).

**From equation 3 (the fundamental form of orbital velocity equation), we get an equation of nearby orbit's Orbital velocity

$$V = [(GM)/R]^{1/2}$$

**And from equation 6 we get another form of equation for orbital speed (when h is negligible) at near earth orbit

$$V_{orbital} = R . (g/r)^{1/2} = R . (g/R)^{1/2} = (gR)^{1/2}$$

Orbital Velocity Formula

Here you get a set of Orbital Velocity Equations or expressions are as follows-

$$V_{orbital} = [(GM)/r]^{1/2}$$

$$V_{orbital} = R \,.\, (g/r)^{1/2}$$

And for Nearby Orbit

$$V_{orbital} = [(GM)/R]^{1/2}$$

$$V_{orbital} = (gR)^{1/2}$$

Orbital Velocity and Radius of the Orbit

From both the equations it's evident that as the radius of the circular path increases, the minimum velocity requirement of the satellite (orbital velocity) to maintain its circular track decreases.

In other words, if the satellite rotates closer to the earth, it has to move faster. If the radius is more then this velocity in orbit is less.

Question :

Calculate the orbital velocity of earth so that the satellite revolves round the earth if radius of earth R = 6.5 × 106 m, mass of earth M = 5.5 × 1024 kg and Gravitational constant G = 6.67 × 10-11 m3/s2 kg.

Solution:

Given:

R = 6.5 × 106 m

M = 5.5 × 1024 kg

G = 6.67 × 10^{-11} m3/s2 kg

The Orbital velocity formula is given by

Vorbit = √GM / R

= √6.67 ×10−11 × 5.5×1024 / 6.5 ×106

= √36.68 x 1013 / 6.5 x 106

= 7.5 km/s

Example :

A satellite launch is made for the study of Jupiter.Determine its velocity so that it orbit round the Jupiter.

Given: Radius of Jupiter R = 70.5 × 106 m

Mass of jupiter M = 1.5 × 1027 Kg

Gravitational constant G = 6.67 × 10^{-11} m^3/s^2 kg

Solution:

When the given parameters are substituted in the orbital velocity formula, we get

Vorbit = √GM / R

= √6.673×10−11×1.5×1027 / 70.5×106

= √10.0095 x 1016 / 70.5 x 106

= √0.141 x 1010

3.754 x 109 m/s.

Patched Conic Approximation

The patched-conic approximation offers an efficient method for describing interplanetary orbits. By partitioning the overall orbit into a series of two-body orbits, it greatly simplifies mission analysis. For each of the portions of an orbit, one gravitational force is assumed to be acting upon the spacecraft at a time. To illustrate the efficiency of the patched-conic approximation, we partition the standard Hohmann transfer of a spacecraft traveling from Earth to Mars into three separate conic stages. During the initial portion of the voyage, we approximate the transfer as a hyperbolic departure orbit with its primary focus positioned at the center of the Earth. After escaping the Earth's sphere of influence, the spacecraft then enters its elliptic orbit about the sun. Following this second stage, the spacecraft enters Mars' sphere of influence. Once again, we approximate the motion as a hyperbolic orbit, this time with its focus located at the center of Mars.

Determining the Heliocentric Departure Velocity

A typical application of the patched-conic solution is to determine the approximate magnitude of Δv needed to complete a certain transfer mission. We first seek the necessary heliocentric velocity v_1 as the spacecraft leaves the Earth's sphere of influence. This particular velocity is illustrated in figure. The v_1 necessary to complete the Hohmann transfer may be computed as

$$v_1 = \sqrt{\frac{2\mu_\odot}{r_\oplus + r_\sigma}\left(\frac{r_\sigma}{r_\oplus}\right)}$$

where $\mu_\odot$ denotes the sun's gravitational coefficient, $r_\oplus$ denotes the Earth's mean orbit radius, and r_σ denotes Mars' mean orbit radius. For a complete derivation of Equation 1, consult Schaub and Junkins. Once the heliocentric departure velocity is calculated, Δv_1 may be computed as

$$\Delta v_1 = v_1 - v_\oplus = v_\oplus \left(\sqrt{\frac{2r_♂}{r_\oplus + r_♂}} - 1 \right)$$

where $v_\oplus$ is the Earth's mean heliocentric velocity, as shown in figure. Because $r_♂ > r_\oplus$, the resulting Δv_1 will be positive.

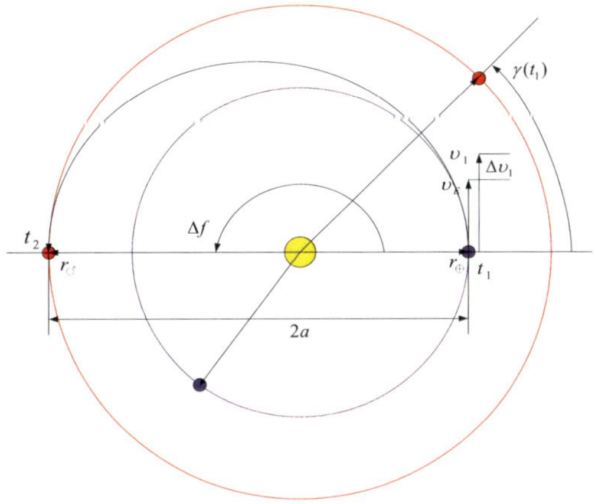

Figure: Illustration of the Hohmann transfer from Earth to Mars

Leaving Earth's Sphere of Influence

The following discussion offers a closer examination of how the spacecraft escapes Earth's sphere of influence. Figure offers an illustration of the departure.

Throughout the following analysis, heliocentric velocities are expressed as v_i, while planet-centric velocities are denoted as v_i. Because the spacecraft is required to converge to some velocity v_1 as it leaves Earth's sphere of influence, the departure orbit must be hyperbolic. The necessary Earth-relative velocity v1 at the limit of the sphere of influence is computed as

$$\text{v}_1 = v_1 - v_\oplus$$

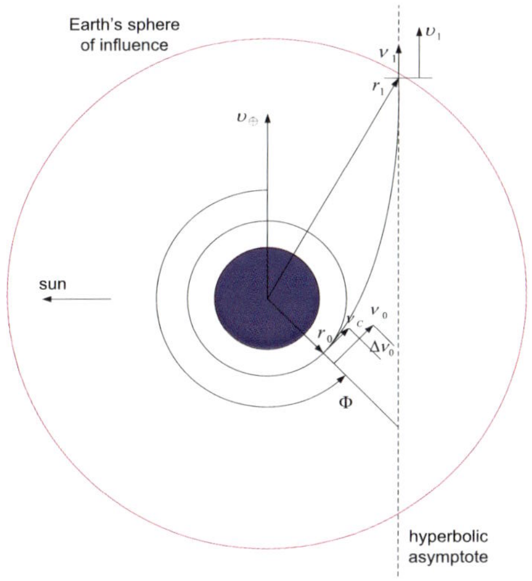

Figure: Illustration of the hyperbolic departure from Earth's sphere of influence

We can also use the vis-viva equation to determine the Earth-relative velocity v_1 as $v_1 = \sqrt{\frac{2\mu_\oplus}{r_1} - \frac{\mu_\oplus}{\alpha_h}} \approx \sqrt{-\frac{\mu_\oplus}{\alpha_h}}$ here $\mu_\oplus$ denotes the Earth's gravitational coefficient and α_h corresponds to the semimajor axis of the departure hyperbola. We approximate $\approx \infty$ due to the assumption that the spacecraft trajectory asymptotically approaches its limiting value at time t_1. Therefore, we can relate the departure hyperbola's semi-major axis to either v_1 or υ_1 via

$$\alpha_h = \frac{-\mu_\oplus}{v_1^2} = -\frac{\mu_\oplus}{(\upsilon_1 - \upsilon_\oplus)^2}$$

Using the vis-viva equation once again, the Earth-relative speed v_0 that the vehicle must have in order to initiate the hyperbolic transfer orbit at t_0 becomes

$$v_0 = \sqrt{\frac{2\mu_\oplus}{r_0} - \frac{\mu_\oplus}{\alpha_h}}$$

where r_0 denotes the spacecraft's initial parking orbit radius about the Earth. After substituting the relation for α_h given in Equation $v_1 = \sqrt{\frac{2\mu_\oplus}{r_1} - \frac{\mu_\oplus}{\alpha_h}} \approx \sqrt{-\frac{\mu_\oplus}{\alpha_h}}$, the speed v_0 is expressed as

$$v_0^2 = v_1^2 + \frac{2\mu_\oplus}{r_0}$$

At this point, it is important to note that once v_1 and r_0 are chosen for a particular mission, the corresponding patched-conic approximation for v_0 is set.

In order to maintain its initial parking orbit about Earth, the spacecraft has a critical speed of

$$v_c = \sqrt{\frac{\mu_\oplus}{r_0}}$$

The initial burn required to begin the hyperbolic transfer is given as

$$\Delta v_0 = v_c - v_c = \sqrt{2v_c^2 + v_1^2} - v_c$$

As shown in figure, the point where the initial Δv_0 burn must be applied is defined via the angle Φ. For any transfer to an outer planet, the spacecraft's velocity should asymptotically align itself with Earth's heliocentric velocity. Thus, the burn angle Φ may be determined from the geometry of the departure hyperbola as

$$\Phi = \cos^{-1}\left(\frac{1}{e_h}\right) + \pi$$

where e_h refers to the eccentricity of the hyperbolic departure orbit. In order to find the departure eccentricity, the orbit's angular momentum must be analyzed.

Entering Mars' Sphere of Influence

We now analyze the patched-conic approximation of how the spacecraft enters Mars' sphere of influence. Figure offers an illustration of the arrival orbit. It is typical for any spacecraft traveling to an outer planet to enter that planet's sphere of influence ahead of the planet. The spacecraft reaches the outer planet at the apoapses of the transfer orbit. Therefore, the spacecraft's speed will be less than that of the planet, allowing the planet to overtake it. Once again, using the vis-viva equation, we find the heliocentric arrival velocity υ_2 of the spacecraft to be

$$\upsilon_2 = \sqrt{2\mu_\odot \left(\frac{1}{r_♂} - \frac{1}{r_\oplus}\right) + \upsilon_1^2}$$

When the spacecraft arrives at Mars, it will most likely cross Mars' sphere of influence with some heading angle σ_2.. For a perfect Hohmann transfer, the value of σ_2 would be exactly equal to 0 degrees. To compute the spacecraft's Mars-centric velocity vector ν_2, Mars' heliocentric velocity must be subracted from that of the spacecraft:

$$\boldsymbol{\nu}_2 = \boldsymbol{\upsilon}_2 - \boldsymbol{\upsilon}_♂$$

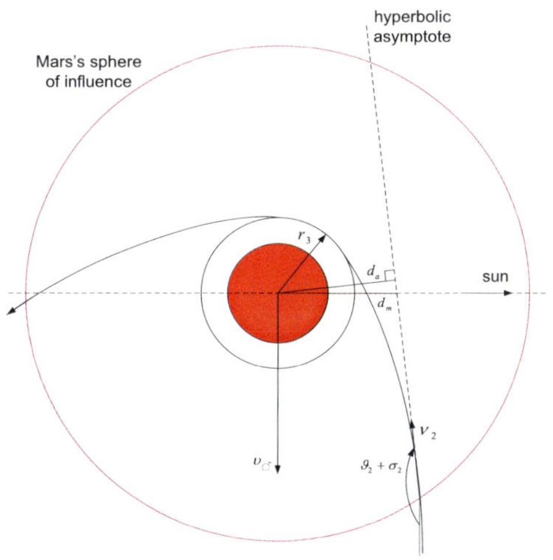

Illustration of the hyperbolic arrival at Mars's sphere of influence

Via the law of cosines, the magnitude of ν_2 is calculated as:

Check the equation

$$\nu_2 = \sqrt{\upsilon_2^2 + \upsilon_♂^2 - 2\upsilon_2\upsilon_♂ \cos\sigma_2}$$

Identical to the process used to describe the departure orbit, we use the energy (visviva) equation to determine the semi-major axis of the arrival orbit through

$$\frac{1}{a_h} = \frac{2}{r_2} - \frac{\nu_2^2}{\mu_♂}$$

Making the patched-conic assumption that the spacecraft's approach orbit is hyperbolic, we approximate a_h as

$$a_h = -\frac{\mu_{♂}}{\nu_2^2}$$

where $r_2 \approx \infty$. If the Hohmann orbit were perfect, the spacecraft would impact the Martian surface. To avoid this occurence, the hyperbolic arrival trajectory is aimed such that it will miss Mars by some miss distance d_m, as shown in figure. However, from the spacecraft's perspective, it is easiest to estimate the shortest distance d_a between the approach asymptote and Mars. Similar to the departure orbit, we examine the spacecraft's constant angular momentum in order to determine the arrival eccentricity e_h. The transfer mission is usually designed in such a way that the periapses radius is equivalent to the final parking orbit radius. Thus, the final orbit radius about Mars is uniquely determined once both the eccentricity e_h and arrival speed v_2 are given. Because e_h depends upon the miss distance, the arrival is actually set with prescribed values of d_m and v_2.

Gravity Assist

The "gravity assist" concept has proven fundamental to exploring our "back yard" the solar system. The technique has even been employed at least once to rescue an Earth-orbiting communications satellite whose launch vehicle failed to place it in its intended geosynchronous orbit.

From Jupiter's point of view, the situation is similar to a bicyclist speeding up going downhill into a valley, then slowing down again on the uphill part of the road.

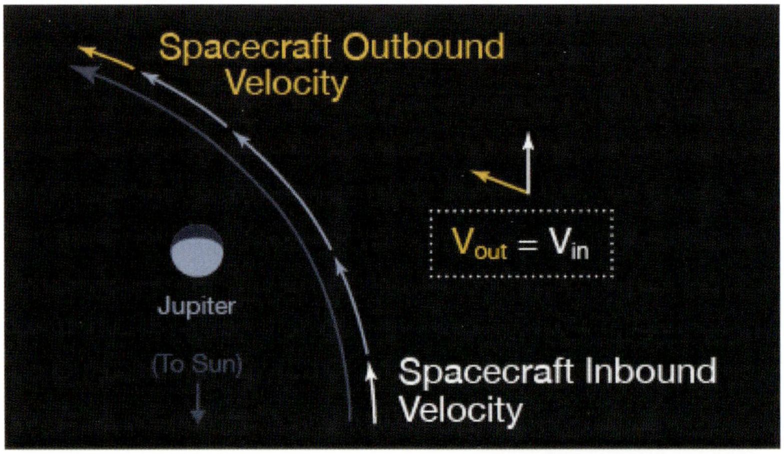

An illustration of the spacecraft's speed relative to Jupiter during a gravity-assist flyby

In the vector diagram at left, the situation is simplified to two dimensions. You can see the magnitude and direction of the spacecraft's velocity on its way in towards Jupiter in the lower right. At the upper left, you can see that the accelerating force of Jupiter's gravitation has made a significant change in the direction of the spacecraft's velocity, but not in its magnitude. (These represent velocity at "infinity," from Jupiter, that is, before and after being noticeably changed by Jupiter's presence.) Near the middle of the diagram, the long arrow shows that there's a significant, but temporary, increase in the magnitude (speed). Note these speeds are all with respect to Jupiter.

To look at the same phenomenon in terms of a cyclist, V_{IN} shows the cyclist approaching a downhill grade into a canyon. V_{OUT} shows that the cyclist slowed down again at the top of the ensuing uphill grade (of course this cycling analogy asks us to ignore air friction and vehicle friction, etc., which are virtually absent in the spacecraft's case). Indeed, after negotiating the canyon, the cyclist's direction has changed, but in the end s/he has not made a lasting change in speed (unless you don't ignore all that friction etc.).

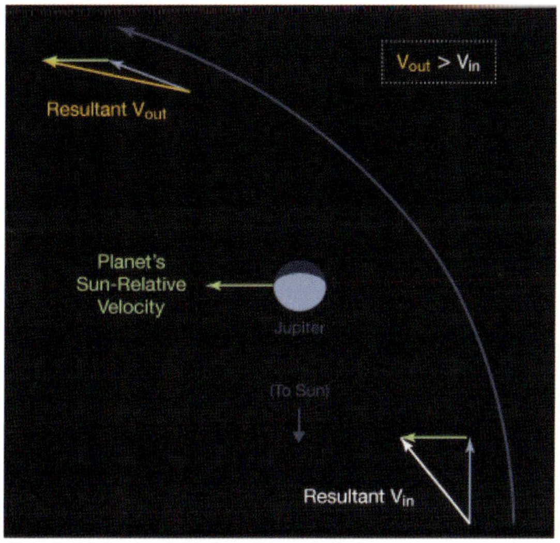

Sun-centric vectors

The planet's own motion is a key. A gravity assist with Jupiter involves not a stationary planet as considered above, but a planet with enormous angular momentum as it revolves around the Sun. In the diagram at right, Jupiter's motion along its solar orbit has been illustrated with a vector colored red (simplified, of course: Jupiter revolves along an arc, not a straight line. Imagine the Sun situated below the bottom of the diagram). The spacecraft acquires this Sun-relative vector, or a significant portion of it, during its interaction with Jupiter.

You can see how the red vector is added to V_{IN} and V_{OUT}. The resulting vector shows how the spacecraft's velocity, relative to the Sun, takes on a nice boost from Jupiter. Notice how rotation of the vector from V_{IN} to V_{OUT} (the bending of the spacecraft's path by the planet's gravity) helps increase the result. This trajectory bending is the other key.

The spacecraft is a physical mass, so it has its own gravitation. That's how the spacecraft can tug on Jupiter and actually decrease the planet's orbital momentum by a tiny amount. In the exchange, the spacecraft acquires momentum from Jupiter a significant amount, compared to the momentum the spacecraft already had.

Purpose

A spacecraft traveling from Earth to an inner planet will increase its relative speed because it is falling toward the Sun, and a spacecraft traveling from Earth to an outer planet will decrease its speed because it is leaving the vicinity of the Sun.

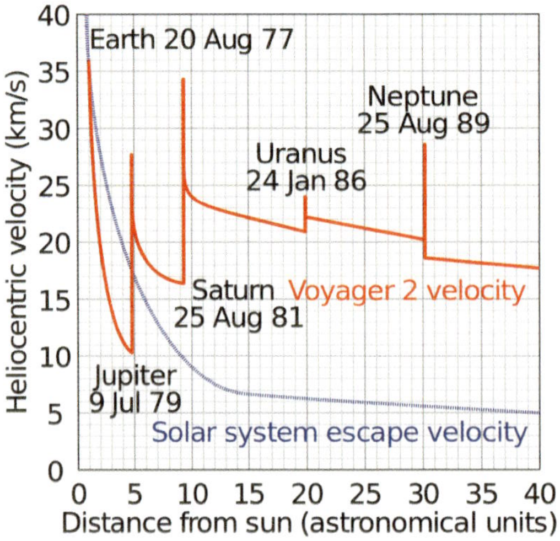

Plot of Voyager 2's heliocentric velocity against its distance from the Sun, illustrating the use of gravity assist to accelerate the spacecraft by Jupiter, Saturn and Uranus. To observe Triton, Voyager 2 passed over Neptune's north pole resulting in an acceleration out of the plane of the ecliptic and reduced velocity away from the Sun

Although the orbital speed of an inner planet is greater than that of the Earth, a spacecraft traveling to an inner planet, even at the minimum speed needed to reach it, is still accelerated by the Sun's gravity to a speed notably greater than the orbital speed of that destination planet. If the spacecraft's purpose is only to fly by the inner planet, then there is typically no need to slow the spacecraft. However, if the spacecraft is to be inserted into orbit about that inner planet, then there must be some way to slow it down.

Similarly, while the orbital speed of an outer planet is less than that of the Earth, a spacecraft leaving the Earth at the minimum speed needed to travel to some outer planet is slowed by the Sun's gravity to a speed far less than the orbital speed of that outer planet. Thus, there must be some way to accelerate the spacecraft when it reaches that outer planet if it is to enter orbit about it. However, if the spacecraft is accelerated to more than the minimum required, less total propellant will be needed to enter orbit about the target planet. In addition, accelerating the spacecraft early in the flight reduces the travel time.

Rocket engines can certainly be used to increase and decrease the speed of the spacecraft. However, rocket thrust takes propellant, propellant has mass, and even a small change in velocity (known as Δv, or "delta-v", the delta symbol being used to represent a change and "v" signifying velocity) translates to a far larger requirement for propellant needed to escape Earth's gravity well. This is because not only must the primary-stage engines lift the extra propellant, they must also lift the extra propellant beyond that, which is needed to lift *that* additional propellant. Thus the liftoff mass requirement increases exponentially with an increase in the required delta-v of the spacecraft.

Because additional fuel is needed to lift fuel into space, space missions are designed with a tight propellant "budget", known as the "delta-v budget". The delta-v budget is in effect the total propellant that will be available after leaving the earth, for speeding up, slowing down, stabilization against external buffeting (by particles or other external effects), or direction changes, if it cannot acquire more propellant. The entire mission must be planned within that capability. Therefore, methods of speed and direction change that do not require fuel to be burned are advantageous, because they allow extra maneuvering capability and course enhancement, without spending fuel from the limited amount which has been carried into space. Gravity assist maneuvers can greatly change the speed of a spacecraft without expending propellant, and can save significant amounts of propellant, so they are a very common technique to save fuel.

Examples:

- The Messenger mission used gravity assist maneuvering to slow the spacecraft on its way to Mercury; however, since Mercury has almost no atmosphere, aerobraking could not be used for insertion into orbit around it.
- Journeys to the nearest planets, Mars and Venus, often use a Hohmann transfer orbit, an elliptical path which starts as a tangent to one planet's orbit round the Sun and finishes as a tangent to the other. When other bodies are unavailable for gravity assists, this often takes the minimum amount of propellant.
- Even using a Hohmann transfer orbit, travel to the outer planets (Jupiter, Saturn, Uranus, and Neptune) would require an extremely large delta-v budget and powerful (or very long-burning) rockets to escape the Sun's gravity, and a very high speed to complete the journey in years rather than decades. Gravitational assist maneuvers offer a way to gain a very high speed without using propellant, therefore as of 2017, all missions to the outer planets have used them.
    - The 1997 *Cassini–Huygens* mission to Saturn is an example of a mission to the outer Solar System. It used repeated gravity assist manouvres - Venus twice, and Earth and Jupiter once each - to travel 3.4 billion km in a little over 6 years, arriving in 2004, which was far faster and more fuel-economical than attempting to travel the "straight line" 1.4 billion km to Saturn directly without gravitational assistance.

Limits

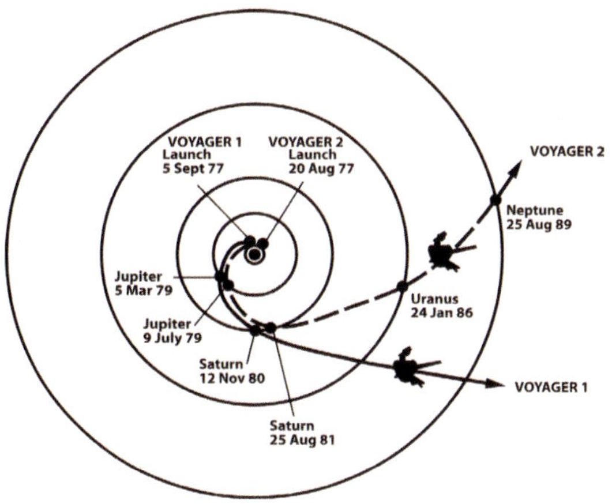

The trajectories that enabled NASA's twin Voyager spacecraft to tour the four giant planets and achieve velocity to escape the Solar System

The main practical limit to the use of a gravity assist maneuver is that planets and other large masses are seldom in the right places to enable a voyage to a particular destination. For example, the Voyager missions which started in the late 1970s were made possible by the "Grand Tour" alignment of Jupiter, Saturn, Uranus and Neptune. A similar alignment will not occur again until the middle of the 22nd century. That is an extreme case, but even for less ambitious missions there are years when the planets are scattered in unsuitable parts of their orbits.

Another limitation is the atmosphere, if any, of the available planet. The closer the spacecraft can approach, the faster its periapsis speed as gravity accelerates the spacecraft, allowing for more kinetic energy to be gained from a rocket burn. However, if a spacecraft gets too deep into the atmosphere, the energy lost to drag can exceed that gained from the planet's gravity. On the other hand, the atmosphere can be used to accomplish aerobraking. There have also been theoretical proposals to use aerodynamic lift as the spacecraft flies through the atmosphere. This maneuver, called an aerogravity assist, could bend the trajectory through a larger angle than gravity alone, and hence increase the gain in energy.

Even in the case of an airless body, there is a limit to how close a spacecraft may approach. The magnitude of the achievable change in velocity depends on the spacecraft's approach velocity and the planet's escape velocity at the point of closest approach (limited by either the surface or the atmosphere.)

Interplanetary slingshots using the Sun itself are not possible because the Sun is at rest relative to the Solar System as a whole. However, thrusting when near the Sun has the same effect as the powered slingshot described as the Oberth effect. This has the potential to magnify a spacecraft's thrusting power enormously, but is limited by the spacecraft's ability to resist the heat.

An interstellar slingshot using the Sun is conceivable, involving for example an object coming from elsewhere in our galaxy and swinging past the Sun to boost its galactic travel. The energy and angular momentum would then come from the Sun's orbit around the Milky Way. This concept features prominently in Arthur C. Clarke's 1972 award-winning novel *Rendezvous With Rama*; his story concerns an interstellar spacecraft that uses the Sun to perform this sort of maneuver, and in the process alarms many nervous humans.

A rotating black hole might provide additional assistance, if its spin axis is aligned the right way. General relativity predicts that a large spinning mass produces frame-dragging close to the object, space itself is dragged around in the direction of the spin. Any ordinary rotating object produces this effect. Although attempts to measure frame dragging about the Sun have produced no clear evidence, experiments performed by Gravity Probe B have detected frame-dragging effects caused by Earth. General relativity predicts that a spinning black hole is surrounded by a region of space, called the ergosphere, within which standing still (with respect to the black hole's spin) is impossible, because space itself is dragged at the speed of light in the same direction as the black hole's spin. The Penrose process may offer a way to gain energy from the ergosphere, although it would require the spaceship to dump some "ballast" into the black hole, and the spaceship would have had to expend energy to carry the "ballast" to the black hole.

The Tisserand Parameter and Gravity Assists

The use of gravity assists is constrained by a conserved quantity called the Tisserand parameter (or invariant.) This is an approximation to the Jacobi constant of the restricted three-body problem. Considering the case a comet orbiting the Sun and the effects a Jupiter encounter would have, Tisserand showed that

$$T_P = \frac{a_J}{a} + 2 \cdot \sqrt{\frac{a}{a_J}(1-e^2)}\cos i$$

will remain constant. (Where a is the comet's semi-major axis, e its eccentricity, i its inclination, and a_J is the semi-major axis of Jupiter. This applies when the comet is sufficiently far from Jupiter to have well-defined orbital elements, and to the extent that Jupiter is much less massive than the Sun and on a circular orbit.

This quantity is conserved for any system of three objects, one of which has negligible mass, and another of which is of intermediate mass and on a circular orbit. For example, the Sun, Earth and a spacecraft, or Saturn, Titan and the Cassini spacecraft (using the semi-major axis of the perturbing body instead of a_J.) This imposes a constraint on how a gravity assist may be used to alter a spacecraft's orbit.

The Tisserand parameter will change if the spacecraft makes a propulsive maneuver or a gravity assist of some fourth object. This is one reason why many spacecraft frequently combine Earth and Venus (or Mars) gravity assists or also perform large deep space maneuvers.

Hohmann Transfer Orbit

A Hohmann Transfer is a two-impulse elliptical transfer between two co-planar circular orbits. The transfer itself consists of an elliptical orbit with a perigee at the inner orbit and an apogee at the outer orbit. The fundamental assumption behind the Hohmann transfer, is that there is only one body which exerts a gravitational force on the body of interest, such as a satellite. This is a good model for transferring an earth-based satellite from a low orbit to say a geosynchronous orbit. Inherent in the model is that there is no additional body sharing the orbit which could induce a gravitational attraction on the body of interest. Thus, as we shall discuss, the Hohmann transfer is a good model for the "outer" trajectory of an earth-Mars transfer, but we must pay some attention to "escaping" the earth's gravitational field before we're on our way.

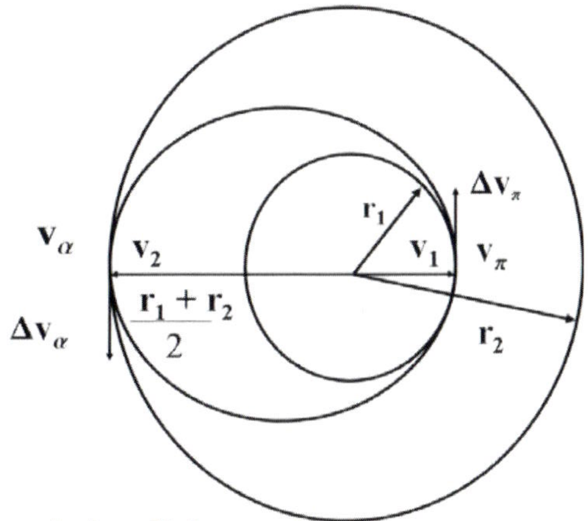

Outer solution: Hohmann transfer

It turns out that this transfer is usually optimal, as it requires the minimum $\Delta v_T = |\Delta v_\pi| + |\Delta v_a|$ to perform a transfer between two circular orbits. The exception for which Hohmann transfers are not optimal is for very large ratios of r_2 / r_1, as discussed below.

$$a = \frac{r_2 + r_1}{2}.$$

Hence, the energy of the transfer orbit is greater than the energy of the inner orbit $(a = r_1)$, and smaller than the energy of the outer orbit $(a = r_2)$. The velocities of the transfer orbit at perigee and apogee are given, from the conservation of energy equation, as

$$v_\pi^2 = \mu\left(\frac{2}{r_1} - \frac{2}{r_1 + r_2}\right)$$

$$v_\alpha^2 = \mu\left(\frac{2}{r_2} - \frac{2}{r_1 + r_2}\right).$$

The velocities of the circular orbits are $v_{c1} = \sqrt{\mu / r_1}$ and $v_{c2} = \sqrt{\mu / r_2}$. Hence, the required impulses at perigee and apogee are,

$$\Delta v_\pi = v_\pi - v_{c1} = \sqrt{\frac{\mu}{r_1}}\left(\sqrt{\frac{2r_2}{r_1 + r_2}} - 1\right)$$

$$\Delta v_a = v_{c2} - v_a = \sqrt{\frac{\mu}{r_2}}\left(1 - \sqrt{\frac{2r_1}{r_1 + r_2}}\right)$$

If the initial orbit has a radius larger than the final orbit, the same strategy can be followed but in this case, negative impulses will be required, first at apogee and then at perigee, to decelerate the satellite.

Example

A communication satellite was carried by the Space Shuttle into low earth orbit (LEO) at an altitude of 322 km and is to be transferred to a geostationary orbit (GEO) at 35,860 km using a Hohmann transfer. The characteristics of the transfer ellipse and the total Δv required, Δv_T, can be determined as follows: For the inner orbit, we have,

$$r_1 = R + d_1 = 6378 \times 10^6 + 322 \times 10^3 = 6.70 \times 10^6 \text{ m}$$

$$v_{c1} = \sqrt{\frac{\mu}{r_1}} = \sqrt{\frac{gR^2}{r_1}} = 7713 \text{ m/s}$$

$$E_1 = -\frac{\mu}{2r_1} = -\frac{gR^2}{2r_1} = -2.975 \times 10^7 \text{ m}^2/\text{s}^2 \text{ (J/kg)}.$$

Similarly, for the outer circular orbit,

$$r_2 = R + d_2 = 42.24 \times 10^6 \text{ m}$$

$$v_{c2} = 3072 \text{ m/s}$$

$$E_2 = -4.718 \times 10^6 \text{ m}^2/\text{s}^2 \text{ (J/kg)}.$$

For the transfer trajectory,

$$2a = r_1 + r_2 = 48.94 \times 10^6 \text{ m}$$

$$E = -\frac{\mu}{2a} = -8.144 \times 10^6 \text{ m}^2/\text{s}^2 \text{ (J/kg)}.$$

which shows that $E_1 < E < E_2$. The velocity of the elliptical transfer orbit at the perigee and apogee can be determined from equations 3 and 4, as,

$$v_\pi = 10{,}130 \text{ m/s},$$

$$v_a = 1{,}067 \text{ m/s}.$$

Since the velocity at the perigee is orthogonal to the position vector, the specific angular momentum of the transfer orbit is,

$$h = r_1 v_\pi = 6.787 \times 10^{10} \text{ m}^2/\text{s},$$

and the eccentricity can be determined as,

$$e = \sqrt{1 + \frac{2Eh^2}{\mu^2}} = 0.7265.$$

Finally, the impulses required are

$$\Delta v_\pi = v_\pi - v_{c1} = 10{,}130 - 7713 = 2414 \text{ m/s}$$

$$\Delta v_a = v_{c2} - v_a = 3072 - 1607 = 1465 \text{ m/s},$$

The sign of the Δv's indicates the direction of thrusting (whether the energy is to be increased or decreased) and the total Δv is the sum of the magnitudes. Thus,

$$\Delta v_T = |\Delta v_\pi| + |\Delta v_a| = 2417 + 1465 = 3882 \text{ m/s}.$$

Since the transfer trajectory is one half of an ellipse, the time of flight (TOF) is simply half of the period,

$$TOF = \pi \sqrt{\frac{(24.47 \times 10^6)^3}{3.986 \times 10^{14}}} = 19{,}050 \text{ s} = 5.29 \text{ h}.$$

In order to illustrate the optimal nature of the Hohmann transfer, we consider now an alternative transfer in which we arbitrarily double the value of the semi-major axis of the Hohmann transfer ellipse, and find the characteristics and Δv_T of the resulting fast transfer. The semi-major axis of the transfer ellipse will be $2a = 98 \times 10^6$ m, and $E = -\mu/(2a) = -4.067 \times 10^6 \text{ m}^2/\text{s}^2$ (J/kg). The velocity of the transfer orbit at departure will be

$$v_\pi = \sqrt{2\left(-4.067 \times 10^6 + \frac{3.986 \times 10^{14}}{6.70 \times 10^6}\right)} = 10.530 \text{ m/s}$$

and,

$$\Delta v_{\pi} = 10{,}530 - 7713 = 2817 \text{ m/s}.$$

The specific angular momentum is,

$$h = 10{,}530 \times \left(6.70 \times 10^{6}\right) = 7.055 \times 10^{10} \text{ m}^2/\text{s},$$

and the orbit's eccentricity, e, is 0.863.

We can now calculate, from the energy conservation equation, the velocity of the transfer orbit at the point of interception with the outer orbit, v_{int},

$$v_{int} = \sqrt{2\left(-4.067 \times 10^{6} + \frac{3.986 \times 10^{14}}{42.24 \times 10^{6}}\right)} = 3277 \text{ m/s}.$$

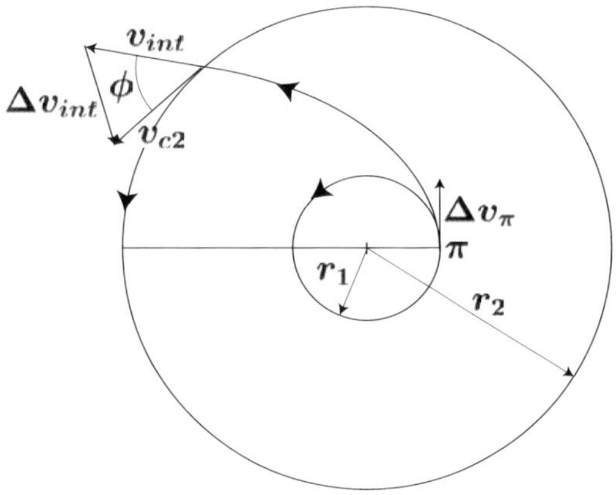

Since the angular momentum, h, is conserved, we can determine the component of v_{int} in the circumferential direction

$$\left(v_{int}\right)_{\theta} = \frac{h}{r^2} = 1670 \text{ m/s}$$

and the elevation angle, ϕ, is thus,

$$\phi = \cos^{-1}\frac{\left(v_{int}\right)_{\theta}}{v_{int}} = 59.36^{o}$$

Finally, from geometrical considerations,

$$\Delta v^2_{int} = v^2_{c2} + v^2_{int} - 2v_{c2}\, v_{int} \cos\phi,$$

which yields $\Delta v_{int} = 3142 \text{ m/s}$, so that

$$\Delta_{vT} = 2817 + 3142 = 5959 \text{ m/s}.$$

Comparing to the value of the Hohmann transfer Δ_{vT} of 3875 m/s, we see that the fast transfer requires a Δ_{vT} which is 54% higher. The analysis of elliptical trajectories which intersect the circular orbits at an angle is referred to as Lambert's problem. These trajectories can have faster transit times but at a greater cost in energy.

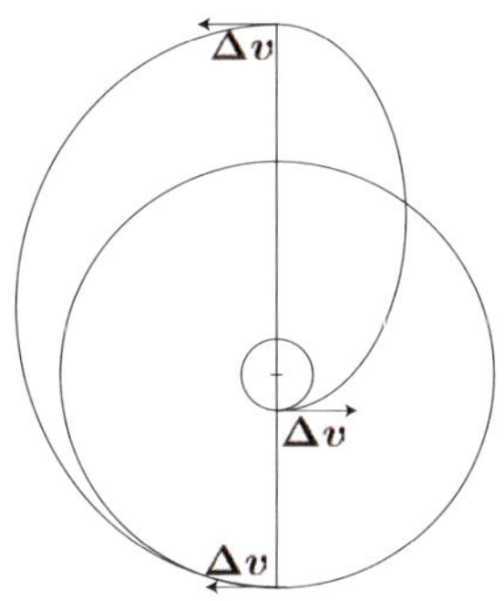

It can be shown that when the separation between the inner and outer orbits is very large ($r_2 > 11.9r_1$) (a situation which rarely occurs), a three impulse transfer comprising of two ellipses can be more energy efficient than a two-impulse Hohmann transfer. This transfer is illustrated in the picture below. Notice that the distance from the origin at which the two transfer ellipses intersect is a free parameter, which can be determined to minimize the total Δ_v. Notice also that the final impulse is a Δ_v which opposes direction of motion, in order to decelerate from the large energy ellipse to the final circular orbit. Although this transfer may be more energy efficient relative to the two-impulse Hohmann transfer, it often involves much larger travel times. Try it!

Example

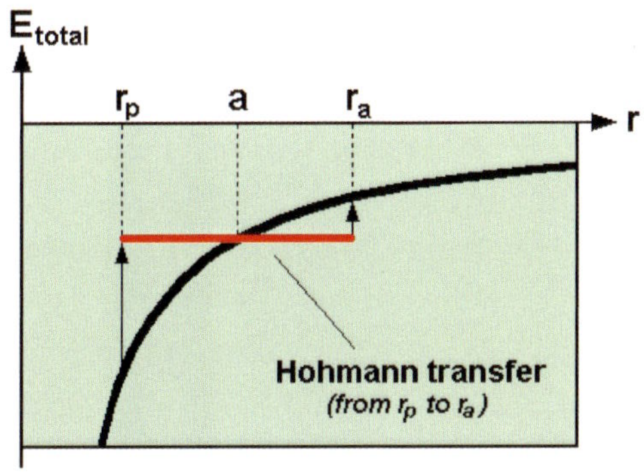

Total energy balance during a Hohmann transfer between two circular orbits with first radius r_p and second radius r_a

Consider a geostationary transfer orbit, beginning at r_1 = 6,678 km (altitude 300 km) and ending in a geostationary orbit with r_2 = 42,164 km (altitude 35,786 km).

In the smaller circular orbit the speed is 7.73 km/s; in the larger one, 3.07 km/s. In the elliptical orbit in between the speed varies from 10.15 km/s at the perigee to 1.61 km/s at the apogee.

The Δv for the two burns are thus 10.15 – 7.73 = 2.42 and 3.07 – 1.61 = 1.46 km/s, together 3.88 km/s.

This is *greater* than the Δv required for an escape orbit: 10.93 – 7.73 = 3.20 km/s. Applying a Δv at the Low Earth orbit (LEO) of only 0.78 km/s more (3.20–2.42) would give the rocket the escape speed, which is less than the Δv of 1.46 km/s required to circularize the geosynchronous orbit. This illustrates the Oberth effect that at large speeds the same Δv provides more specific orbital energy, and energy increase is maximized if one spends the Δv as quickly as possible, rather than spending some, being decelerated by gravity, and then spending some more to overcome the deceleration (of course, the objective of a Hohmann transfer orbit is different).

Worst Case, Maximum Delta-v

As the example above demonstrates, the Δv required to perform a Hohmann transfer between two circular orbits is not the greatest when the destination radius is infinite. (Escape speed is $\sqrt{2}$ times orbital speed, so the Δv required to escape is $\sqrt{2} - 1$ (41.4%) of the orbital speed.) The Δv required is greatest (53.0% of smaller orbital speed) when the radius of the larger orbit is 15.5817 times that of the smaller orbit. This number is the positive root of $x^3 - 15x^2 - 9x - 1 = 0$, which is $5+4\sqrt{7}\cos\left(\frac{1}{3}\arctan\frac{\sqrt{3}}{37}\right)$. For higher orbit ratios the Δv required for the second burn decreases faster than the first increases.

Application to Interplanetary Travel

When used to move a spacecraft from orbiting one planet to orbiting another, the situation becomes somewhat more complex, but much less delta-v is required, due to the Oberth effect, than the sum of the delta-v required to escape the first planet plus the delta-v required for a Hohmann transfer to the second planet.

For example, consider a spacecraft travelling from the Earth to Mars. At the beginning of its journey, the spacecraft will already have a certain velocity and kinetic energy associated with its orbit around Earth. During the burn the rocket engine applies its delta-v, but the kinetic energy increases as a square law, until it is sufficient to escape the planet's gravitational potential, and then burns more so as to gain enough energy to get into the Hohmann transfer orbit (around the Sun). Because the rocket engine is able to make use of the initial kinetic energy of the propellant, far less delta-v is required over and above that needed to reach escape velocity, and the optimum situation is when the transfer burn is made at minimum altitude (low periapsis) above the planet. The delta-v needed is only 3.6 km/s, only about 0.4 km/s more than needed to escape Earth, even though this results in the spacecraft going 2.9 km/s faster than the Earth as it heads off for Mars.

At the other end, the spacecraft will need a certain velocity to orbit Mars, which will actually be less than the velocity needed to continue orbiting the Sun in the transfer orbit, let alone attempting to orbit the Sun in a Mars-like orbit. Therefore, the spacecraft will have to decelerate in order for the gravity of Mars to capture it. This capture burn should optimally be done at low altitude to also make best use of Oberth effect. Therefore, relatively small amounts of thrust at either end of the trip are needed to arrange the transfer compared to the free space situation.

However, with any Hohmann transfer, the alignment of the two planets in their orbits is crucial – the destination planet and the spacecraft must arrive at the same point in their respective orbits around the Sun at the same time. This requirement for alignment gives rise to the concept of launch windows.

The term lunar transfer orbit (LTO) is used for the moon.

It is possible to apply the formula given above to calculate the Δv in km/s needed to enter a Hohmann transfer orbit to arrive at various destinations from Earth (assuming circular orbits for the planets). In this table, the column labeled "Δv to enter Hohmann orbit from Earth's orbit" gives the change from Earth's velocity to the velocity needed to get on a Hohmann ellipse whose other end will be at the desired distance from the sun. The column labeled "v exiting LEO" gives the velocity needed (in a non-rotating frame of reference centered on the earth) when 300 km above the Earth's surface. This is obtained by adding to the specific kinetic energy the square of the speed (7.73 km/s) of this low Earth orbit (that is, the depth of Earth's gravity well at this LEO). The column "Δv from LEO" is simply the previous speed minus 7.73 km/s.

Destination	Orbital radius (AU)	Δv (km/s)		
		to enter Hohmann orbit from Earth's orbit	exiting LEO	from LEO
Sun	0	29.8	31.7	24.0
Mercury	0.39	7.5	13.3	5.5
Venus	0.72	2.5	11.2	3.5
Mars	1.52	2.9	11.3	3.6
Jupiter	5.2	8.8	14.0	6.3
Saturn	9.54	10.3	15.0	7.3
Uranus	19.19	11.3	15.7	8.0
Neptune	30.07	11.7	16.0	8.2
Pluto	39.48	11.8	16.1	8.4
Infinity	∞	12.3	16.5	8.8

Note that in most cases, Δv from LEO is less than the Δv to enter Hohmann orbit from Earth's orbit.

To get to the sun, it is actually not necessary to use a Δv of 24 km/s. One can use 8.8 km/s to go very far away from the sun, then use a negligible Δv to bring the angular momentum to zero, and then fall into the sun. This can be considered a sequence of two Hohmann transfers, one up and one down. Also, the table does not give the values that would apply when using the moon for a gravity assist. There are also possibilities of using one planet, like Venus which is the easiest to get to, to assist getting to other planets or the sun.

Hohmann Transfer versus Low Thrust Orbits

Low-thrust Transfer

Low-thrust engines can perform an approximation of a Hohmann transfer orbit, by creating a gradual enlargement of the initial circular orbit through carefully timed engine firings. This requires a change in velocity (delta-v) that is greater than the two-impulse transfer orbit and takes longer to complete.

Engines such as ion thrusters are more difficult to analyze with the delta-v model. These engines offer a very low thrust and at the same time, much higher delta-v budget, much higher specific impulse, lower mass of fuel and engine. A 2-burn Hohmann transfer maneuver would be impractical with such a low thrust; the maneuver mainly optimizes the use of fuel, but in this situation there is relatively plenty of it.

If only low-thrust maneuvers are planned on a mission, then continuously firing a low-thrust, but very high-efficiency engine might generate a higher delta-v and at the same time use less propellant than a rocket engine.

Going from one circular orbit to another by gradually changing the radius simply requires the same delta-v as the difference between the two speeds. Such maneuver requires more delta-v than a 2-burn Hohmann transfer maneuver, but does so with continuous low thrust rather than the short applications of high thrust.

The amount of propellant mass used measures the efficiency of the maneuver plus the hardware employed for it. The total delta-v used measures the efficiency of the maneuver only. For electric propulsion systems, which tend to be low-thrust, the high efficiency of the propulsive system usually compensates for the higher delta-V compared to the more efficient Hohmann maneuver.

Transfer orbit using electrical propulsion or low-thrust engines optimize the transfer time to reach the final orbit and not the delta-v as in the Hohmann transfer orbit. For geostationary orbit, the initial orbit is set to be supersynchronous and by thrusting continuously in the direction of the velocity at apogee, the transfer orbit transforms to a circular geosynchronous one. This method however takes much longer to achieve due to the low thrust injected into the orbit.

Interplanetary Transport Network

In 1997, a set of orbits known as the Interplanetary Transport Network (ITN) was published, providing even lower propulsive delta-v (though much slower and longer) paths between different orbits than Hohmann transfer orbits. The Interplanetary Transport Network is different in nature than Hohmann transfers because Hohmann transfers assume only one large body whereas the Interplanetary Transport Network does not. The Interplanetary Transport Network is able to achieve the use of less propulsive delta-v by employing gravity assist from the planets.

Geostationary Transfer Orbit

A geosynchronous orbit is a high Earth orbit that allows satellites to match Earth's rotation. Located at 22,236 miles (35,786 kilometers) above Earth's equator, this position is a valuable spot for monitoring weather, communications and surveillance.

"Because the satellite orbits at the same speed that the Earth is turning, the satellite seems to stay in place over a single longitude, though it may drift north to south," NASA wrote on its Earth Observatory website.

Satellites are designed to orbit Earth in one of three basic orbits defined by their distance from the planet: low Earth orbit, medium Earth orbit or high Earth orbit. The higher a satellite is above

Earth (or any other world for that matter), the slower it moves. This is because of the effect of Earth's gravity; it pulls more strongly at satellites that are closer to its center than satellites that are farther away.

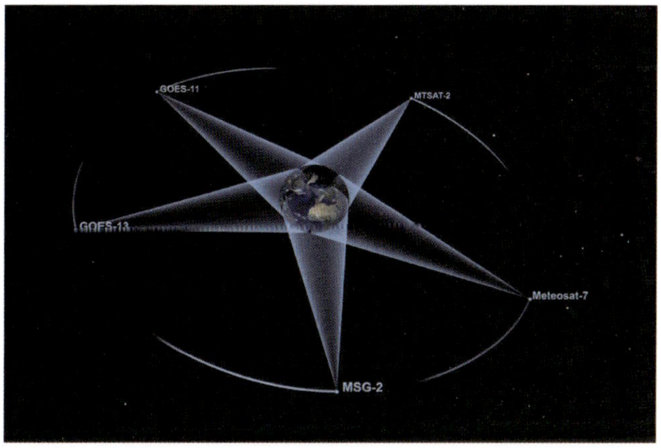

The National Oceanic and Atmospheric Administration's Climate Prediction Center (CPC) uses five geosynchronous satellites: GOES-11, GOES-13, MSG-2, Meteosat-7 and MTSAT-2

So a satellite at low Earth orbit such as the International Space Station, at roughly 250 miles (400 km) will move over the surface, seeing different regions at different times of day. Those at medium Earth orbit (between about 2,000 and 35,780 km, or 1,242 and 22,232 miles) move more slowly, allowing for more detailed studies of a region. At geosynchronous orbit, however, the orbital period of the satellite matches the orbit of the Earth (roughly 24 hours), and the satellite appears virtually still over one spot; it stays at the same longitude, but its orbit may be tilted, or inclined, a few degrees north or south.

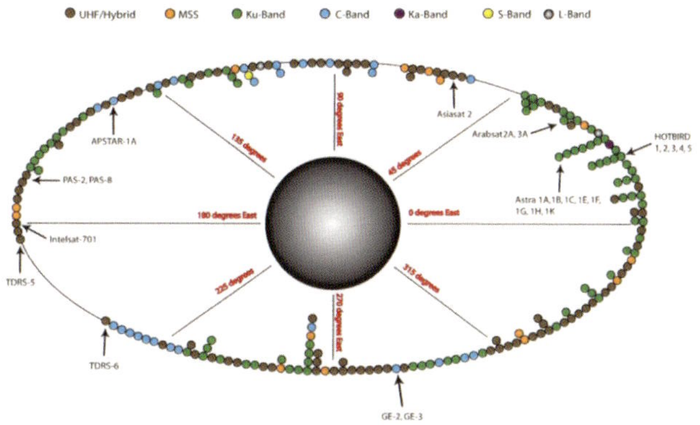

This image depicts the geostationary equatorial orbit in which most communications and weather satellites are located

Benefits

A satellite in geosynchronous orbit can see one spot of the planet almost all of the time. For Earth observation, this allows the satellite to look at how much a region changes over months or years. The drawback is the satellite is limited to a small parcel of ground; if a natural disaster happens elsewhere, for example, the satellite won't be able to move there due to fuel requirements.

This is a large benefit for the military. If, for example, the United States is concerned about activities in a certain region of the world or it wants to see how its troops are doing a geosynchronous orbit allows constant pictures and other surveillance of one particular region. An example of this is the United States' Wideband Global SATCOM 5, which launched in 2013. Joining a "constellation" of four other WGS satellites, it extends the military's communications system to provide blanket coverage over virtually the entire planet. The network serves troops, ships, drones and civilian leaders and is supposed to provide communications for ground personnel.

Communications for civilians also benefit from geosynchronous orbit. There are numerous companies that provide telephone, Internet, television and other services from satellites in that orbital slot. Because the satellite is constantly hovering over one spot on the ground, communications from that location are reliable as long as the satellite is well connected to the location you want to communicate with.

Orbital Competition

According to Satellite Signals, there are 402 satellites in geosynchronous orbit. At geosynchronous orbit, the "ring" around Earth can accommodate a number of satellites 1,800 altogether, according to one analysis by Lawrence Roberts. However, there are obvious space and technological limitations.

Specifically, satellites must remain in a very confined area and not drift too far from their assigned "slot" above Earth; otherwise they may pose a threat to other satellites. The International Telecommunication Union assigns slots for geosynchronous orbit and settles disputes between countries about slots.

Similarly, it is considered good practice to move almost-dead satellites into a "graveyard" orbit above geosynchronous orbit before they run out of fuel, to clear the way for the next generation.

The satellites must also be located far enough away from each other so their communications don't interfere with each other, which could mean a separation of anything between 1 and 3 degrees. As technology has improved, it's possible to pack more satellites into a smaller spot.

References

- Curtis, Howard D., (2009). Orbital Mechanics for Engineering Students, 2e. New York: Elsevier. ISBN 978-0-12-374778-5
- Keplers-laws-of-planetary-motion, science: britannica.com, Retrieved 22 May 2018
- Jonathan McDowell, "Kick In the Apogee: 40 years of upper stage applications for solid rocket motors, 1957-1997", 33rd AIAA Joint Propulsion Conference, July 4, 1997. abstract. Retrieved 18 July 2017
- Egorov, Vsevolod Alexandrovich (September 1957). "Specific problems of a flight to the moon". Physics-Uspekhi. 63 (9): 73–117. doi:10.3367/UFNr.0063.195709f.0073
- Orbital-velocity-satellite: physicsteacher.in, Retrieved 29 June 2018
- Bate, Roger R.; Mueller, Donald D.; White, Jerry E. (1971). Fundamentals of Astrodynamics. New York: Dover Publications. ISBN 0-486-60061-0
- National Earth Science Teachers Association (9 October 2008). "Data Table for Planets and Dwarf Planets". Windows to the Universe. Retrieved 2 August 2018
- Geosynchronous-orbit-29222: space.com, Retrieved 22 July 2018

- Everitt, C. W. F.; et al. (June 2011). "Gravity Probe B: Final Results of a Space Experiment to Test General Relativity". Physical Review Letters. 106 (22). 221101. arXiv:1105.3456. Bibcode:2011PhRvL.106v1101E. doi:10.1103/PhysRevLett.106.221101. PMID 21702590
- Satellite-communication-orbital-mechanics, tutorialspoint.com, Retrieved 31 March 2018
- Sellers, Jerry J.; Astore, William J.; Giffen, Robert B.; Larson, Wiley J. (2004). Kirkpatrick, Douglas H., ed. Understanding Space: An Introduction to Astronautics (2 ed.). McGraw Hill. p. 228. ISBN 0-07-242468-0

3

Radiation and Matter Interactions

An understanding of the interactions between matter and radiation can be developed from the study of microwave, infrared, visible, ultraviolet and X-ray interactions with matter as well as from the analysis of how photons, electrons and positrons interact with matter. This chapter has been carefully written to provide a comprehensive understanding of all aspects of interactions between matter and radiation.

Microwave Interactions

The quantum energy of microwave photons is in the range 0.00001 to 0.001 eV which is in the range of energies separating the quantum states of molecular rotation and torsion. The interaction of microwaves with matter other than metallic conductors will be to rotate molecules and produce heat as result of that molecular motion. Conductors will strongly absorb microwaves and any lower frequencies because they will cause electric currents which will heat the material. Most matter, including the human body, is largely transparent to microwaves. High intensity microwaves, as in a microwave oven where they pass back and forth through the food millions of times, will heat the material by producing molecular rotations and torsions. Since the quantum energies are a million times lower than those of x-rays, they cannot produce ionization and the characteristic types of radiation damage associated with ionizing radiation.

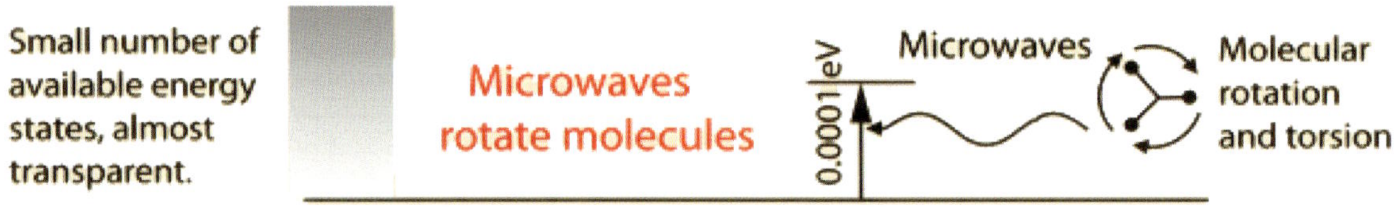

Infrared Interactions

The quantum energy of infrared photons is in the range 0.001 to 1.7 eV which is in the range of energies separating the quantum states of molecular vibrations. Infrared is absorbed more strongly than microwaves, but less strongly than visible light. The result of infrared absorption is heating of the tissue since it increases molecular vibrational activity. Infrared radiation does penetrate the skin further than visible light and can thus be used for photographic imaging of subcutaneous blood vessels.

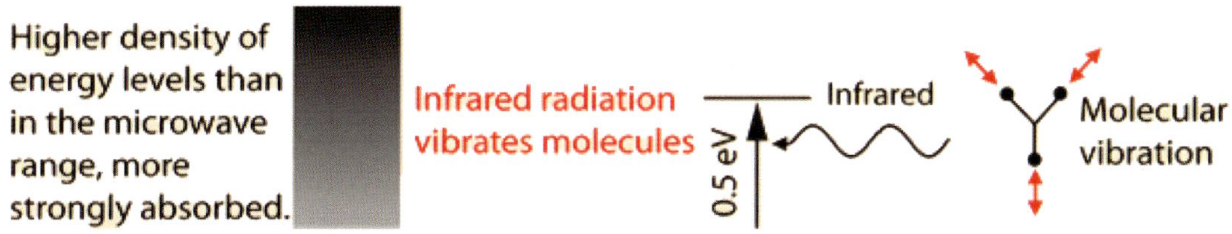

Visible Light Interactions

The primary mechanism for the absorption of visible light photons is the elevation of electrons to higher energy levels. There are many available states, so visible light is absorbed strongly. With a strong light source, red light can be transmitted through the hand or a fold of skin, showing that the red end of the spectrum is not absorbed as strongly as the violet end.

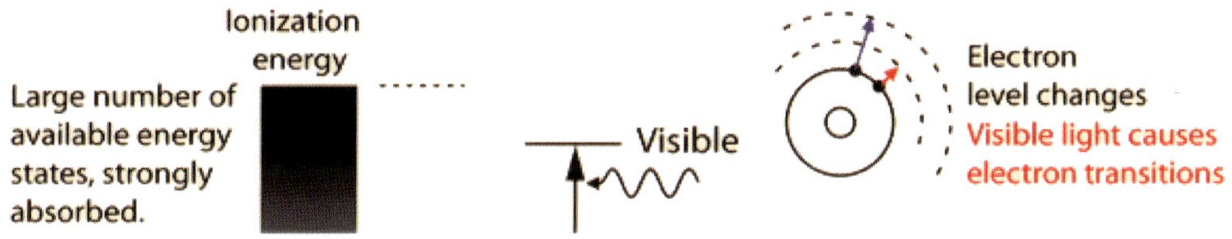

While exposure to visible light causes heating, it does not cause ionization with its risks. You may be heated by the sun through a car windshield, but you will not be sunburned - that is an effect of the higher frequency uv part of sunlight which is blocked by the glass of the windshield.

Ultraviolet Interactions

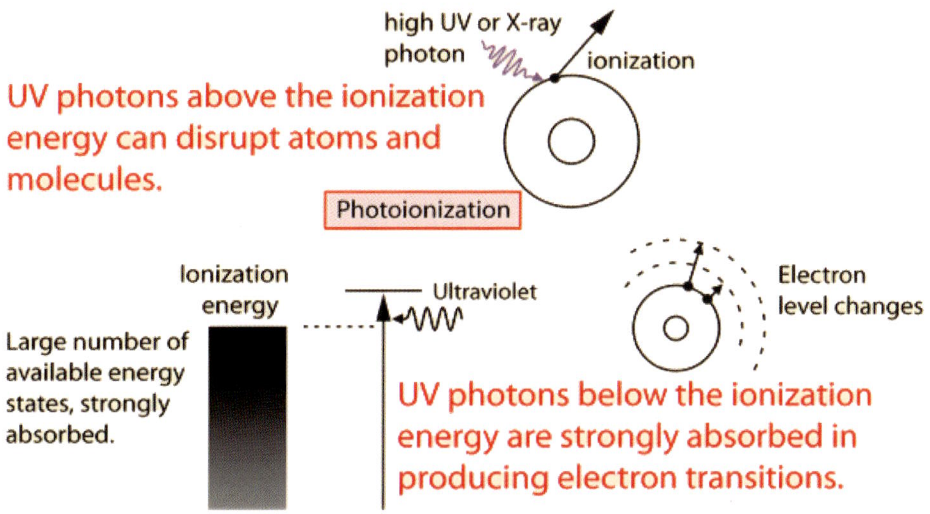

The near ultraviolet is absorbed very strongly in the surface layer of the skin by electron transitions. As you go to higher energies, the ionization energies for many molecules are reached and the more dangerous photoionization processes take place. Sunburn is primarily an effect of uv, and ionization produces the risk of skin cancer.

The ozone layer in the upper atmosphere is important for human health because it absorbs most of the harmful ultraviolet radiation from the sun before it reaches the surface. The higher frequencies in the ultraviolet are ionizing radiation and can produce harmful physiological effects ranging from sunburn to skin cancer.

Health concerns for UV exposure are mostly for the range 290-330 nm in wavelength, the range called UVB. According to Scotto, the most effective biological wavelength for producing skin burns is 297 nm. Their research indicates that the biological effects increase logarithmically within the UVB range, with 330 nm being only 0.1% as effective as 297 nm for biological effects. So it is clearly important to control exposure to UVB.

X-rays Interactions

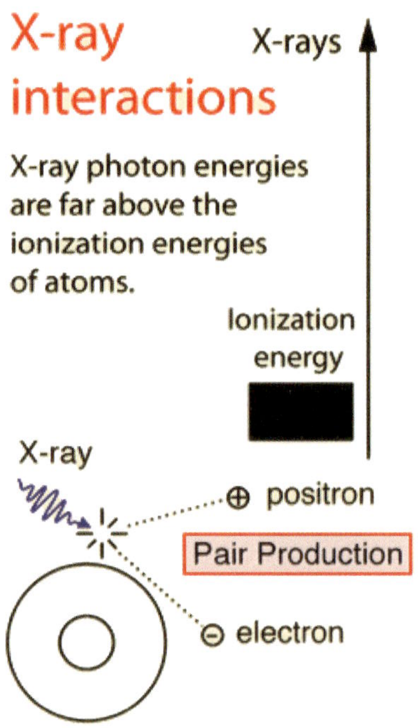

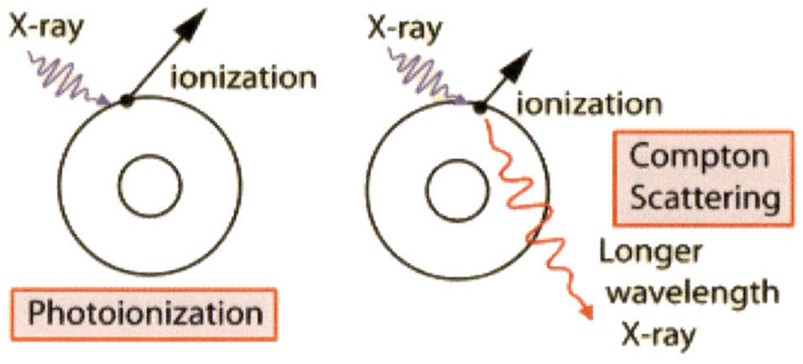

Since the quantum energies of x-ray photons are much too high to be absorbed in electron transitions between states for most atoms, they can interact with an electron only by knocking it completely out of the atom. That is, all x-rays are classified as ionizing radiation. This can occur by giving all of the energy to an electron (photoionization) or by giving part of the energy to the electron and the remainder to a lower energy photon (Compton scattering). At sufficiently high energies, the x-ray photon can create an electron positron pair.

Photon Interactions

Photons are quanta of electromagnetic energy that exhibit both wave-like and particle-like properties. A photon can be considered to have a wavelength and frequency (like a wave), as well as momentum and energy (like a particle). Despite carrying electromagnetic energy, a photon has no 'charge' and has a much lower chance of interacting with matter than charged particles such as electrons and protons.

There are six ways in which photons may interact with matter:

- Coherent Scattering
- Photoelectric Effect
- Incoherent Scattering, also known as Compton Scattering or Compton Effect
- Pair Production
- Triplet Production
- Photodisintegration

These may cause the photon to attenuate (lose some of its energy and/or disappear). Photon interactions are very important when considering how a photon beam interacts with a patient.

Coherent Scattering - σ_{coh}

Coherent (or Rayleigh) scattering occurs at low photon energies. A photon may interact with an orbital electron and is then deflected (or scattered) at a small angle. There is no change in energy of the photon and no other effects occur.

Coherent scattering is more probably as atomic number increases and as photon energy decreases. This makes it of some concern in diagnostic x-ray, where it can cause loss of contrast and blurring. For radiotherapy it has minimal impact on attenuation.

Photoelectric Effect - τ

The photoelectric effect occurs when a photon interacts with an orbital electron whose binding energy is close to that of the photon energy. In this scenario, the photon disappears and all of its energy is given to the orbital electron, which is then ejected from the atom with kinetic energy equivalent to the photon energy minus the binding energy.

The space left by the departing electron is filled by another orbital electron, with emission of characteristic x-rays or Auger electrons (when a characteristic x-ray causes ejection of another orbital electron within the same atom). In soft tissues, these characteristic x-rays have a low energy (about 500 eV) and are absorbed locally.

The probability of the photoelectric effect occurring is:

- Inversely proportional to the cube of the photon energy (ie: $\frac{1}{E^3}$)
- Proportional the cube of the atomic number i.e.: Z^3

Therefore, as photon energy increases, the likelihood of the photoelectric effect decreases rapidly.

As the atomic number of a material increases, the likelihood of the photoelectric effect increases rapidly.

The photoelectric effect ceases to become likely at energies over 100-200 keV in water, but remains an important factor for attenuation in lead at energies up to 1 MeV.

Absorption Edges

An important concept is that when the photon energy is very closely aligned with the binding energy of a particular electron shell, the likelihood of photoelectric interactions jumps. This is an absorption edge, and reflects the ability of photons above that energy to interact more frequently.

Incoherent Scattering (Compton Effect) - σ_{inc}

Incoherent scattering is the most important interaction in radiotherapy. It occurs when a photon has a much greater amount of energy than the binding energy of the electron, effectively considering the electron as 'free'. In this interaction, the photon interacts with the 'free' electron, giving up some of its energy and undergoing scattering. The electron receives the energy and is set in motion in a different direction. Photons may be scattered in any direction, whereas electrons are only scattered in a forward direction (maximum 90° from the original path of the photon). The energy given to the electron is dependent on the photon energy (low energies tend to give up minimal energy to the electron) and the angle of impact.

- Direct impact scatters the photon at 180 degrees and gives a large amount of energy to the electron which continuous in the photons original path.
- Glancing hits cause no photon scattering and give a small amount of energy to the electron, which moves at 90 degrees away from the photon direction.

Importantly, incoherent scattering is not directly related to atomic number but rather the concentration of electrons in tissue. As atomic number rises, the density of electrons falls slowly, so incoherent scattering becomes slightly less likely in high Z materials. Incoherent scattering also decreases with energy, but nowhere near as rapidly as the photoelectric effect. This makes it more relevant as photon energies rise above the K-shell binding energies of orbital electrons.

Pair Production - κ

Pair production occurs when a photon passes very close to the nucleus of an atom. If the energy of the photon is high enough, the photon may disappear and 'create' an electron and a positron. The new particles move away with the remaining energy of the photon converted to kinetic energy. The positron will typically annihilate once it loses its kinetic energy, giving rise to annihilation quanta (two photons with energy of 0.511 keV).

Situations in which Pair Production is Possible

Pair production only occurs when photon energy is high enough and when there is an object of sufficient mass to take on the momentum gained by the new particles.

- Photon energy must be at least 1.022 MeV to generate a positron/electron pair. This is because 1.022 MeV is required to create the mass of the electron and the positron (through $E = mc^2$.

- Pair production only occurs in close proximity to the nucleus. This is so momentum of the electron and positron is conserved.
    - It is possible for pair production to occur in proximity to an electron. This is called triplet production, below.

Relation of Pair Production with Atomic Number and Photon Energy

Pair production does not occur with photon energy under 1.022 MeV. Once this threshold is reached, it becomes more likely as photon energy increases.

Pair production is related to the atomic number of a material through Z^2.

Triplet Production - κ_{tr}

Triplet production is a special case of pair production which occurs in the vicinity of an orbital electron. The photon disappears and the energy is used to create an electron and positron. The orbital electron also receives energy and is freed from the atom. The threshold for this to occur is a photon of 2.044 MeV.

Photodisintegration - π

Photodisintegration is an uncommon event that occurs when a photon is absorbed by the nucleus of an atom. The photon is destroyed and a nucleon (either a proton or a neutron) is released. The threshold for this effect is over 10 MeV for most nuclei (with the exception of beryllium and deuterium, where it is 2 MeV). Even at high energies, photodisintegration is an uncommon event and does not attenuate a substantial portion of a photon beam. It is more important for radiation protection concerns - if neutrons are released they are highly penetrating and can convert atoms into unstable isotopes. The release of a nucleon from the atom in question also usually results in a radioactive daughter product. The production of neutrons in high energy linear accelerators means that bunkers must be regularly ventilated to prevent buildup of radioactive gasses.

Electron Interactions

Electrons accelerated onto a material result in a number of interactions with the atoms of the target sample. Accelerated electrons can pass through the sample without interaction, undergo elastic scattering and can be inelastically scattered. Elastic and inelastic scattering result in a number of signals that are used for imaging, quantitative and semi-quantitative information of the target sample and generation of an X-ray source. Typical signals used for imaging include secondary electrons (SE), backscattered electrons (BSE), cathodoluminescence (CL), auger electrons and characteristic X-rays. Quantitative and semiquantitative analyses of materials as well as element mapping typically utilize characteristic X-rays. Bremsstrahlung (continuum) radiation is a continuous spectrum of X-rays from zero to the energy of the electron beam, and forms a background in which characteristic X-ray must be considered. Further, X-rays generated from a specific target material are used as the roughly fixed-wavelength energy source for X-ray diffraction (XRD) and X-ray fluorescence (XRF) investigations.

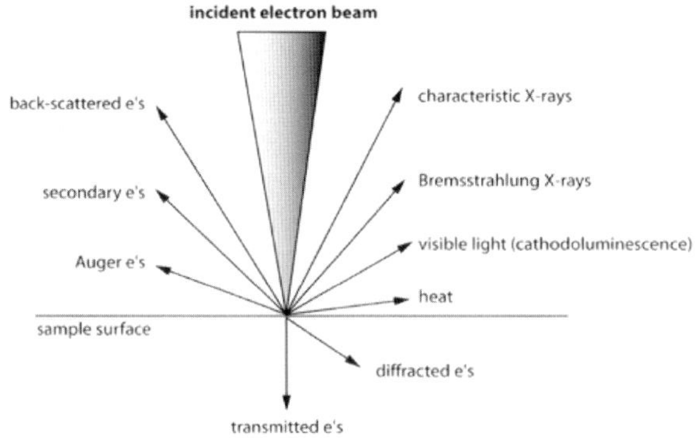

Figure: Types of interactions between electrons and a sample

Where an electron beam impinges on a sample, electron scattering and photon- and X-ray-production develops in a volume (the electron interaction volume) that is dependent on several factors as shown in figure below. These include:

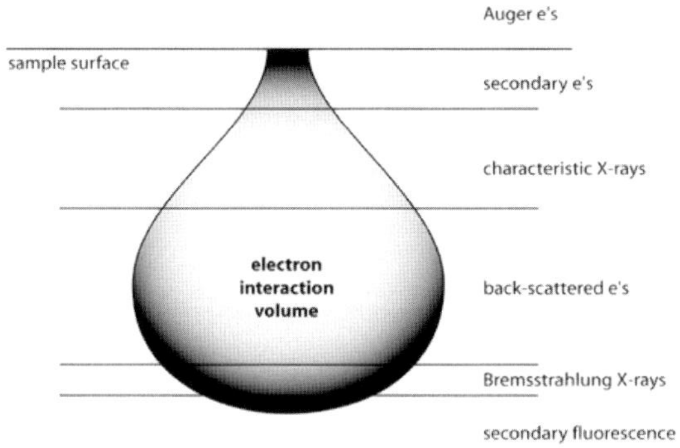

Figure: Electron interaction volume within a sample

Each of the electron, X-ray and light emissions generated after the impingement of a high-energy electron beam, develops at different depths (volumes) within the sample. In turn, the volume of generation of the emitted energy controls the spatial resolution of the signal.

- The energy of the incident beam (accelerating potential) increases the interaction volume, but decreases the elastic scattering (e.g. backscattering).
- The interaction volume decreases as a function of the mean atomic weight.
- Smaller and more asymmetric interaction volumes develop in samples tilted relative to the impinging electron beam.

Each of the signals used for imaging or X-ray generation is generated from different electron interaction volumes and, in turn, each of the signals has different imaging or analytical resolution. Auger and Secondary images have the best imaging resolution, being generated in the smallest volume near the surface of the sample. Backscattered electrons are generated over a larger volume resulting in images of intermediate resolution. Cathodoluminescence is generated over the largest volume, even larger than Bremsstrahlung radiation, resulting in images with the poorest resolution.

Positron Interactions

The coulomb forces that constitute the major mechanism of energy loss for electrons are present for either positive or negative charge on the particle and constitute the major mechanism of energy loss also for positrons. Whatever the interaction involves a repulsive or attractive force between the incident particle and orbital electron (or atomic nucleus), the impulse and energy transfer for particles of equal mass are about the same. Therefore positrons interact similarly with matter when they are energetic. The track of positrons in material is similar to the track of electrons. Even their specific energy loss and range are about the same for equal initial energies.

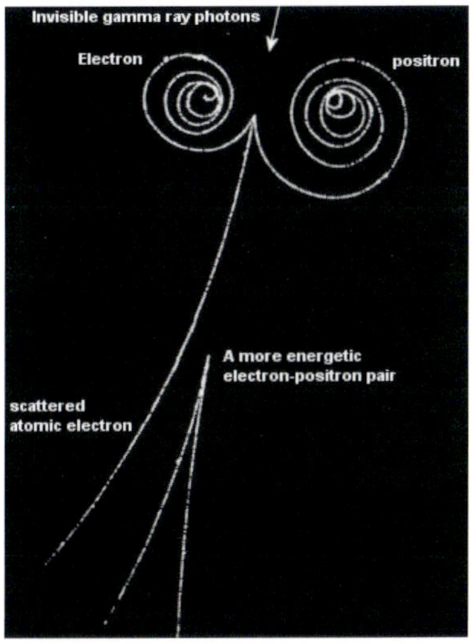

At the end of their path, positrons differ significantly from electrons. When a positron (antimatter particle) comes to rest, it interacts with an electron (matter particle), resulting in the annihilation of the both particles and the complete conversion of their rest mass to pure energy (according to the $E=mc^2$ formula) in the form of two oppositely directed 0.511 MeV gamma rays (photons).

Positron Annihilation

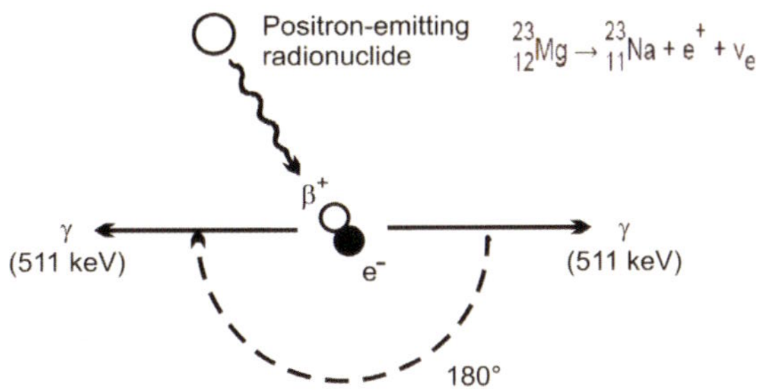

When a positron (antimatter particle) comes to rest, it interacts with an electron, resulting in the annihilation of the both particles and the complete conversion of their rest mass to pure energy in the form of two oppositely directed 0.511 MeV photons.

Electron–positron annihilation occurs when a negatively charged electron and a positively charged positron collide.When a low-energy electron annihilates a low-energy positron (antiparticle of electron), they can only produce two or more photons (gamma rays). The production of only one photon is forbidden because of conservation of linear momentum and total energy. The production of another particle is also forbidden because of both particles (electron-positron) together do not carry enough mass-energy to produce heavier particles. When an electron and a positron collide, they annihilate resulting in the complete conversion of their rest mass to pure energy (according to the $E=mc^2$ formula) in the form of two oppositely directed 0.511 MeV gamma rays (photons).

$$e^- + e^+ \rightarrow \gamma + \gamma \text{ (2x 0.511 MeV)}$$

This process must satisfy a number of conservation laws, including:

- Conservation of electric charge. The net charge before and after is zero.
- Conservation of linear momentum and total energy. T
- Conservation of angular momentum.

4

The Sun

The sun is the center of the solar system. It is a sphere of hot plasma. The aim of this chapter is to provide an overview of the Sun. It includes detailed analysis of the solar core, corona, photosphere, chromosphere, solar cycle and solar dynamo.

Sun, star around which Earth and the other components of the solar system revolve. It is the dominant body of the system, constituting more than 99 percent of its entire mass. The Sun is the source of an enormous amount of energy, a portion of which provides Earth with the light and heat necessary to support life.

Joshua trees at sunset, Joshua Tree National Park

The Sun is classified as a G2 V star, with G2 standing for the second hottest stars of the yellow G class of surface temperature about 5,800 kelvins (K) and the V representing a main sequence, or dwarf, star, the typical star for this temperature class. (G stars are so called because of the prominence of a band of atomic and molecular spectral lines that the German physicist Joseph von Fraunhofer designated G.) The Sun exists in the outer part of the Milky Way Galaxy and was formed from material that had been processed inside a supernova. The Sun is not, as is often said, a small star. Although it falls midway between the biggest and smallest stars of its type, there are so many dwarf stars that the Sun falls in the top 5 percent of stars in the neighbourhood that immediately surrounds it.

Physical Properties

The radius of the Sun, $R_{\odot}$, is 109 times that of Earth, but its distance from Earth is 215 $R_{\odot}$, so it subtends an angle of only $^{1}/_{2}{}^{\circ}$ in the sky, roughly the same as that of the Moon. By comparison, Proxima Centauri, the next closest star to Earth, is 250,000 times farther away, and its relative apparent brightness is reduced by the square of that ratio, or 62 billion times. The temperature of the Sun's surface is so high that no solid or liquid can exist there; the constituent materials are predominantly gaseous atoms, with a very small number of molecules. As a result, there is no fixed surface. The surface viewed from Earth, called the photosphere, is the layer from which most of the radiation reaches us; the radiation from below is absorbed and reradiated, and the emission from overlying layers drops sharply, by about a factor of six every 200 kilometres (124 miles). The Sun is so far from Earth that this slightly fuzzy surface cannot be resolved, and so the limb (the visible edge) appears sharp.

The eight planets of the solar system and Pluto, in a montage of images scaled to show the approximate sizes of the bodies relative to one another.

Outward from the Sun, which is represented to scale by the yellow segment at the extreme left, are the four rocky terrestrial planets (Mercury, Venus, Earth, and Mars), the four hydrogen-rich giant planets (Jupiter, Saturn, Uranus, and Neptune), and icy, comparatively tiny Pluto.*NASA/Lunar and Planetary Laboratory*

Inland Sea Sunset on the Inland Sea, with the Seto Great Bridge in the foreground.*Tomas Tam*

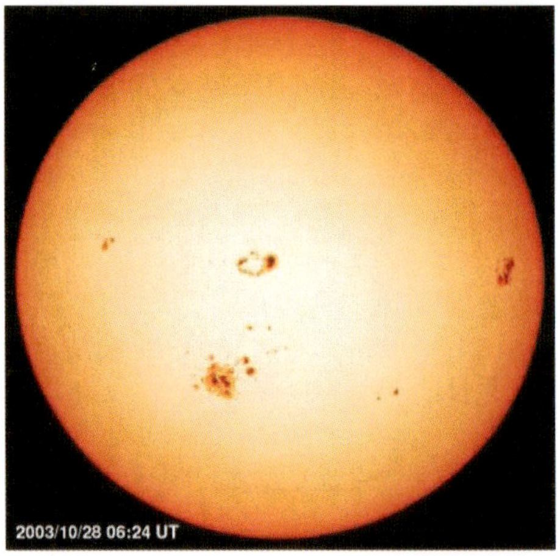

Photosphere of the Sun with sunspots, image

The mass of the Sun, $M_{\odot}$, is 743 times the total mass of all the planets in the solar system and 330,000 times that of Earth. All the interesting planetary and interplanetary gravitational phenomena are negligible effects in comparison to the force exerted by the Sun. Under the force of gravity, the great mass of the Sun presses inward, and to keep the star from collapsing, the central pressure outward must be great enough to support its weight. The density at the Sun's core is about 100 times that of water (roughly six times that at the centre of Earth), but the temperature is at least 15,000,000 K, so the central pressure is at least 10,000 times greater than that at the centre of Earth, which is 3,500 kilobars. The nuclei of atoms are completely stripped of their electrons, and at this high temperature they collide to produce the nuclear reactions that are responsible for generating the energy vital to life on Earth.

While the temperature of the Sun drops from 15,000,000 K at the centre to 5,800 K at the photosphere, a surprising reversal occurs above that point; the temperature drops to a minimum of 4,000 K, then begins to rise in the chromosphere, a layer about 7,000 kilometres high at a temperature of 8,000 K. During a total eclipse the chromosphere appears as a pink ring. Above the chromosphere is a dim, extended halo called the corona, which has a temperature of 1,000,000 K and reaches far past the planets. Beyond a distance of $5R_{\odot}$ from the Sun, the corona flows outward at a speed (near Earth) of 400 kilometres per second (km/s); this flow of charged particles is called the solar wind.

The Sun is a very stable source of energy; its radiative output, called the solar constant, is 1.366 kilowatts per square metre at Earth and varies by no more than 0.1 percent. Superposed on this stable star, however, is an interesting 11-year cycle of magnetic activity manifested by regions of transient strong magnetic fields called sunspots.

The Sun and Magnetism

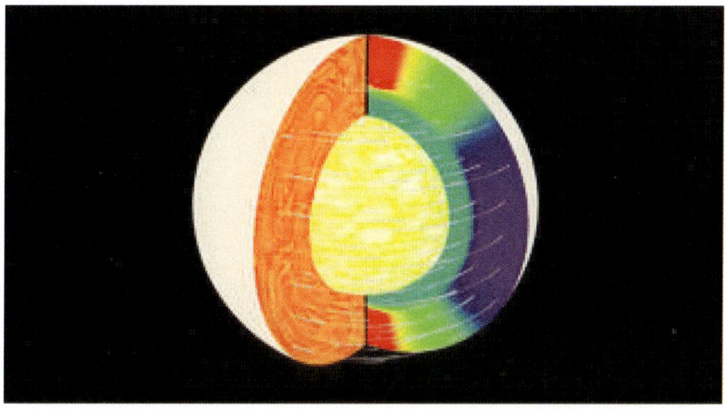

An illustration of the interior plasma flows. The right-hand cross-section is the toroidal flow. The left-hand cross section represents the meridional flow.

It is widely believed that the Sun's magnetic field is generated by electrical currents acting as a magnetic dynamo inside the Sun. These electrical currents are generated by the flow of hot, ionized gases in the Sun's convection zone.

We know a lot about the Sun's magnetic dynamo. It has a 22 year cycle. During the first half of the cycle, the Sun's magnetic north pole is in the northern hemisphere while the magnetic south pole is in the southern hemisphere. Right around the peak of the sunspot cycle (solar maximum), the magnetic poles flip or exchange places so that magnetic north is now located in the southern hemisphere. This flip occurs about every 11 years at solar maximum.

The 22 year magnetic cycle greatly influences the most prominent manifestation of the dynamo, sunspots and active regions, which migrate towards the solar equator from high latitudes over the course of the solar 11 year "sunspot cycle". Sunspots and Active Regions are manifestation of the magnetic field generated in the Sun's interior poking through the visible region of the atmosphere. Active regions are responsible for the production of intense and violent energy burst, called flares, and events where very large amounts of hot gas, trapped by the magnetic field of the active region, are released from the Sun's atmosphere and into space, called coronal mass ejections (CMEs).

Solar Core

The solar core is made up of a really hot and dense gas (in the plasma state). The temperature of 15 million kelvins (27 million degrees Fahrenheit) keeps the core at a gaseous state.

The core is where the energy of the Sun is made. The density and temperature are just right for nuclear fusion reactions to take place. These reactions release energy in two forms, light (electromagnetic radiation) and particles (in particular neutrinos). Despite the reactions, the Sun's core is a very dark place!

The sun's core rotates nearly four times faster than the sun's surface, according to new findings by an international team of astronomers. Scientists had assumed the core was rotating like a merry-go-round at about the same speed as the surfa.

"The most likely explanation is that this core rotation is left over from the period when the sun formed, some 4.6 billion years ago

The rotation of the solar core may give a clue to how the sun formed. After the sun formed, the solar wind likely slowed the rotation of the outer part of the sun, he said. The rotation might also impact sunspots, which also rotate, Ulrich said. Sunspots can be enormous; a single sunspot can even be larger than the Earth.

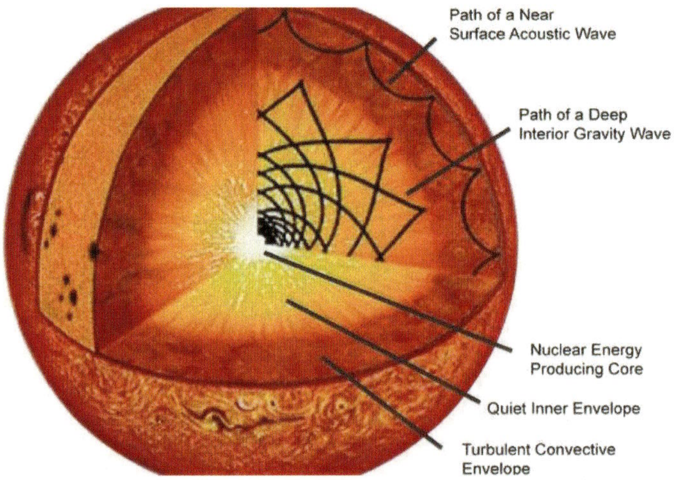

The sun, whose hot, dense core produces nuclear fusion and rotates nearly four times faster than its surface.

The researchers studied surface acoustic waves in the sun's atmosphere, some of which penetrate to the sun's core, where they interact with gravity waves that have a sloshing motion similar to how water would move in a half-filled tanker truck driving on a curvy mountain road. From those observations, they detected the sloshing motions of the solar core. By carefully measuring the acoustic waves, the researchers precisely determined the time it takes an acoustic wave to travel from the surface to the centre of the sun and back again. That travel time turns out to be influenced a slight amount by the sloshing motion of the gravity waves.

The researchers identified the sloshing motion and made the calculations using 16 years of observations from an instrument called GOLF (Global Oscillations at Low Frequency) on a spacecraft called SoHO (the Solar and Heliospheric Observatory) a joint project of the European Space Agency and NASA.

Composition

The Sun at the photosphere is about 73-74% by mass hydrogen, which is the same composition as the atmosphere of Jupiter, and the primordial composition of hydrogen and helium at the earliest star formation after the Big Bang. However, as depth into the Sun increases, fusion decreases the fraction of hydrogen. Traveling inward, hydrogen mass fraction starts to decrease rapidly after the core radius has been reached (it is still about 70% at a radius 25% of the Sun's radius) and inside this, the hydrogen fraction drops rapidly as the core is traversed, until it reaches a low of about 33% hydrogen, at the Sun's centre (radius zero). All but 2% of the remaining plasma mass (i.e., 65%) is helium, at the centre of the Sun.

Radiation Zone

The Sun's radiative zone is the section of the solar interior between the innermost core and the outer convective zone. In the radiative zone, energy generated by nuclear fusion in the core moves outward as electromagnetic radiation. In other words, the energy is conveyed by photons. When the energy reaches the top of the radiative zone, it begins to move in a different fashion in the convective zone. In the convective zone, heat and energy are carried outward along with matter in swirling flows called convection cells. This motion is similar to the roiling flows seen in a pot of boiling water.

The inner parts of the Sun (core and radiative zone) spin differently than the outer layers (convective zone). The boundary between these two types of rotation, which lies between the radiative and convective zones, is called the tachocline.

Many other stars also have radiative zones. The Sun's radiative zone extends from the core outward to about 70% of the Sun's radius. In a smaller (than the Sun) star that is cooler than our Sun, the convective zone tends to be larger, extending deeper into the star's interior. Thus the radiative zone tends to be smaller. In very small, cool stars the convective zone may reach all the way to the star's core, and there may be no radiative zone at all. In a larger (than the Sun) star with a higher temperature, the radiative zone tends to be larger and the convective zone smaller. Especially large, hot stars may not have a convective zone at all - their radiative zone may extend all the way from the core to the star's surface.

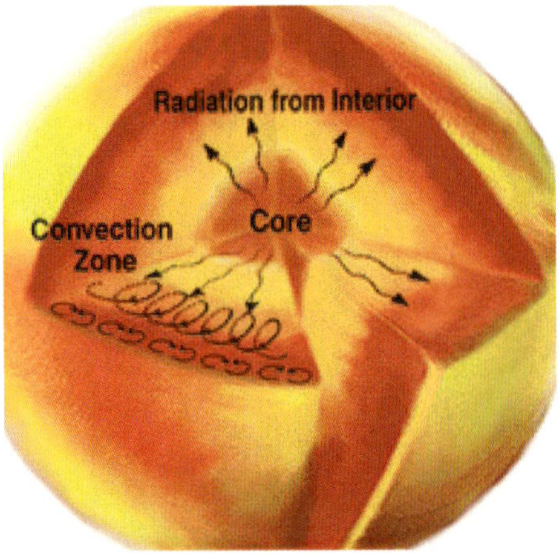

The Sun's radiative zone lies between the incredibly hot core and the outer convective zone.

Convection Zone

In general, convection is another mechanism by which heat can be transferred. In the same manner, the Sun uses this method to transfer the energy it produces from the core towards the outermost region. This process takes place in the region of the Sun called the convection zone. From the core, the energy is transferred into the radiation zone through radiation mechanism. As the energy escape from this region, it begins its journey outward passing through another layer that uses a different means of transfer.

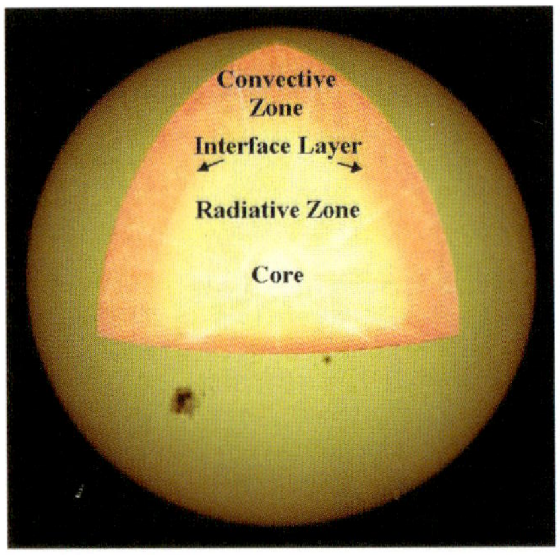

The process occurring in a convection zone is similar to boiling water in a pot thus this is sometimes referred to as boiling zone. Outside the radiation zone, the temperature is much lower compared to the temperature within the region. From about 5 million degrees Kelvin, the temperature drops down to only about 2 million degrees Kelvin. However, this temperature is still relatively higher than within the convection region. What happens during the convection mechanism is that the hotter material near the radiation zone tends to rise. When this reaches the top of the convection region, it cools down and sinks again. This is a direct way to transfer energy and thus occurs much faster than in radiation zone.

The convection zone is considered to be the outermost layer of the interior portion of the sun. This layer extends to about 200,000 km up to the visible surface of the Sun known as the photosphere. When seen through a telescope, a white ball with some dark patches can be observed. These dark patches are termed as sunspots. The entire surface of this region is covered with convection cells similar to what one can be observed on the surface of boiling water. The bright and dark regions represent the hot rising and the cold sinking materials respectively.

Corona

The corona is the outer atmosphere of the Sun. It extends many thousands of kilometers (miles) above the visible "surface" of the Sun, gradually transforming into the solar wind that flows outward through our solar system.

The material in the corona is an extremely hot but very tenuous plasma. The temperature in the corona is more than a million degrees, surprisingly much hotter than the temperature at the Sun's surface which is around 5,500° C (9,940° F or 5,780 kelvins). The pressure and density in the corona is much, much lower than in Earth's atmosphere.

The corona is above the Sun's lower atmosphere, which is called the chromosphere. A relatively narrow area called the transition region separates the corona from the chromosphere. Temperatures rise sharply in the transition region, from thousands of degrees in the chromosphere to more than a million degrees in the corona. The density of plasma falls rapidly through the transition region moving upward from the chromosphere to the corona.

We normally cannot see the solar atmosphere, including the corona. The surface of the Sun is far too bright to allow a glimpse of the much fainter corona. During a total solar eclipse the wispy corona briefly comes into view as the Moon blocks out the solar surface. A special instrument called a coronagraph allows astronomers to view the corona at other times. Some coronagraphs are used with ground-based telescopes; others are carried on satellites.

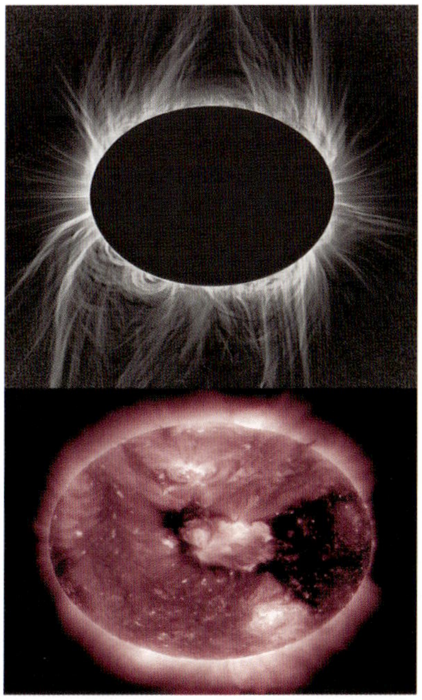

Two views of the Sun's corona: during an eclipse (top) and in ultraviolet light (bottom).

Physical Features

The Sun's corona is much hotter (by a factor of nearly 200) than the visible surface of the Sun: The photosphere's average temperature is 5800 kelvin compared to the corona's one to three million kelvin. The corona is 10^{-12} as dense as the photosphere, however, and so produces about one-millionth as much visible light. The corona is separated from the photosphere by the relatively shallow chromosphere.

The exact mechanism by which the corona is heated is still the subject of some debate. Likely possibilities include induction by the Sun's magnetic field and sonic pressure waves from below (the latter being less probable now that coronae are known to be present in early-type, highly magnetic stars). The outer edges of the Sun's corona are constantly being transported away due to open magnetic flux generating the solar wind.

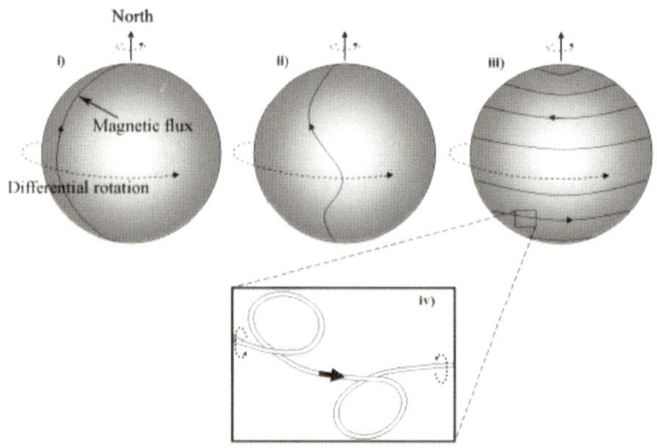

A drawing demonstrating the configuration of solar magnetic flux during the solar cycle.

The Corona is not always evenly distributed across the surface of the sun. During periods of quiet, the corona is more or less confined to the equatorial regions, with coronal holes covering the polar regions. However, during the Sun's active periods, the corona is evenly distributed over the equatorial and polar regions, though it is most prominent in areas with sunspot activity. The solar cycle spans approximately 11 years, from solar minimum to solar maximum, where the solar magnetic field is continually wound up (due to a differential rotation at the solar equator; the equator rotates quicker than the poles). Sunspot activity will be more pronounced at solar maximum where the magnetic field is twisted to a maximum. Associated with sunspots are coronal loops, loops of magnetic flux, upwelling from the solar interior. The magnetic flux pushes the hotter photosphere aside, exposing the cooler plasma below, thus, creating the dark (when compared to the solar disk) spots.

Photosphere

The photosphere is the visible "surface" of the Sun, but is not a true or solid surface because the Sun is completely gaseous. Moving outward from the core of the Sun, the density, temperature, and gas pressure all decrease until, in a thin layer (only 400 kilometers thick), the material gradually changes from being completely opaque (light cannot pass through it) to being completely transparent.

Close inspection of the photosphere shows that it is not a uniformly bright layer, but has variations. It has a mottled or granular appearance, granulation, due to small regions of warmer and cooler material. The brighter regions are about 100 K hotter than surrounding regions and are rising at 2–3 km/s. These regions last only a few minutes before they dissipate and are replaced with other rising masses of warmer gas. Meanwhile, the surrounding cooler and darker regions are sinking back into the solar interior. This is direct observational evidence that the photosphere is also the top of a convective zone in the Sun (like the mass motion in a pot of boiling water that carries heat from the bottom to the top of the water). In the deeper interior, photons are able to move energy outward by slow diffusion in a radiative zone. Convection is far more efficient at moving energy, with energy passing outward through the solar convective zone (the outer 15% of the Sun) in a matter of hours. Photons, however, endure enormous numbers of collisions with nuclei and electrons and may take tens of thousands of years to work their way into the outer regions of the Sun. Larger areas of the photosphere may appear slightly brighter (warmer) or fainter (cooler) than other regions. This supergranulation indicates the existence of larger convective regions.

Within this layer also occur the sunspots, the most visible of photospheric features. Sunspots are temporarily cooler regions appearing as dark spots on the surface of the Sun. Sunspots appear dark only in contrast to surrounding hotter regions. In their centres, the temperature is lower by about 1,200 K, and hence their surface brightness is only one-third that of nearby regions. A typical sunspot appears to have a dark center (umbra) and lighter outer annulus (penumbra) with radial striations caused by the magnetic field associated with the sunspot material. Direct measurement of surface magnetic fields shows that sunspot magnetism is stronger than the surroundings. Sunspots tend to appear in groups (up to 200,000 kilometers in diameter), usually developing rapidly in pairs, then fading away slowly over a few weeks. The leading sunspot in a pair has a magnetic polarity opposite to the trailing spot, with Southern Hemisphere pairs showing the north-south polarity opposite to that of the Northern Hemisphere pairs.

Convection is a turbulent process and produces faculae, short-lived hot jets of material that spurt outward into the layers outside the photosphere. Flares and eruptions, or brightenings of large regions lasting from minutes to hours, are associated with sunspots and again are a result of the convection. Such flares spew out large quantities of charged particles which, if they encounter the Earth four days later, cause distortions in the Earth's magnetic field (a geomagnetic disturbance).

Temperature

The temperature of the photosphere is around 5,500° C (about 9,900° F). The photosphere is much cooler than the Sun's core, which has a temperature well above 10 million degrees. Surprisingly, the photosphere is also much cooler than the Sun's atmosphere above it, which has regions with temperatures ranging up to a couple of million degrees. There is some variation in temperature from place to place in the photosphere - for example, sunspots are relatively cool regions with temperatures as low as 3,000 to 4,500° C (5,400 to 8,100° F).

Location

The area of the Sun's interior immediately below the photosphere is called the convection (or convective) zone. The lower section of the Sun's atmosphere, the chromosphere, lies above the photosphere.

Opacity

The gases that make up the photosphere are completely opaque, meaning that you cannot see through them. Therefore, stating that the sun has "surface" is a misnomer, for the photosphere is not solid.

Composition

The photosphere is constructed of convection cells called granules, which are cells of hot gas 1,000 km in diameter. Each granule lives 8 to 9 minutes, which produces a "boiling" effect.

Chromosphere

The lower region of the Sun's atmosphere is called the chromosphere. Its name comes from the Greek root chroma (meaning color), for it appears bright red when viewed during a solar eclipse.

The chromosphere extends for about 2,000 kilometers (1,200 miles) above the visible surface of the Sun. Compared to the radius of the Sun, which is nearly 700 thousand kilometers (more than 400 thousand miles), the chromosphere is a very thin layer.

The chromosphere is above the photosphere, the visible "surface" of the Sun. It lies below the solar corona, the Sun's upper atmosphere, which extends many thousands of kilometers above the chromosphere into space.

The plasma (electrically charged gas) in the chromosphere has a very low density. It is about ten thousand times less dense than the underlying photosphere, and more than a million times less dense than Earth's atmosphere. Because it is such a thin layer and made of such tenuous plasma, the chromosphere is normally hidden from our view it's light is drowned out by the brilliance of the underlying photosphere. It is typically only seen during a solar eclipse or through the use of special telescopes on satellites.

Composition and Characteristics of the Chromosphere

The flash spectrum has been a valuable tool in the study of the chromosphere. This spectrum is obtained before a solar eclipse reaches totality and is formed from the thin arc of the sun disappearing behind the moon's disk. An analysis of the emission lines gives information about the heights of the chromosphere and the heights at which various elements exist in it. Using the flash spectrum, scientists have found that the chromosphere is composed primarily of hydrogen, which causes its visible pinkish tint, and of sodium, magnesium, helium, calcium, and iron in lesser amounts. The chromosphere consists of three distinct layers that, moving outward from the sun's surface, decrease in density and increase abruptly in temperature. The lower chromosphere is about 10,800°F (6,000°C), the middle rises to 90,000°F (50,000°C), and the upper part, merging into the lower corona, reaches 1,800,000°F (1,000,000°C).

Solar Cycle

Our Sun is a huge ball of electrically-charged hot gas. This charged gas moves, generating a powerful magnetic field. The Sun's magnetic field goes through a cycle, called the solar cycle.

Every 11 years or so, the Sun's magnetic field completely flips. This means that the Sun's north and south poles switch places. Then it takes about another 11 years for the Sun's north and south poles to flip back again.

The solar cycle affects activity on the surface of the Sun, such as sunspots which are caused by the Sun's magnetic fields. As the magnetic fields change, so does the amount of activity on the Sun's surface.

One way to track the solar cycle is by counting the number of sunspots. The beginning of a solar cycle is a solar minimum, or when the Sun has the least sunspots. Over time, solar activity and the number of sunspots increases.

The middle of the solar cycle is the solar maximum, or when the Sun has the most sunspots. As the cycle ends, it fades back to the solar minimum and then a new cycle begins.

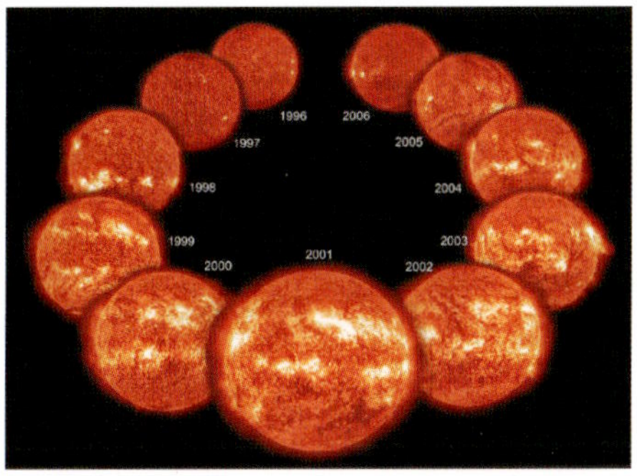

Images of the Sun during one solar cycle. Solar maximum occurred during 2001 while 1996 and 2006 were near solar minimum.

Giant eruptions on the Sun, such as solar flares and coronal mass ejections, also increase during the solar cycle. These eruptions send powerful bursts of energy and material into space.

This activity can have effects on Earth. For example, eruptions can cause lights in the sky, called aurora, or impact radio communications. Extreme eruptions can even affect electricity grids on Earth.

An image of a coronal mass ejection observed by NASA's Solar and Heliospheric Observatory, or SOHO, satellite in 2001.

Some cycles have maximums with lots of sunspots and activity. Other cycles can have very few sunspots and little activity. Scientists work hard to improve our ability to predict the strength and duration of solar cycles. These predictions can help them forecast these solar conditions, called space weather.

Forecasting of the solar cycle can help scientists protect our radio communications on Earth, and help keep NASA satellites and astronauts safe, too.

NASA Astronaut Tim Kopra on a 2015 spacewalk outside the International Space Station

Solar activity can affect satellite electronics and limit their lifetime. Radiation can be dangerous for astronauts who do work on the outside of the International Space Station. If scientists predict an active time in the solar cycle, satellites can be put into safe mode and astronauts can delay their spacewalks.

The Solar Activity Cycle

The solar dynamo yields the theoretical explanation of the basic solar activity cycle which is mainly (but not exclusively) manifested in the periodic appearance of sunspots. Its mean period is 11 yrs (i.e. 22 yrs for the cycle time of the full magnetic oscillation) but this value varies between 8 and 14 yrs. The variability of the activity period can be expressed by the quality $\omega cyc/\delta\omega cyc$ which for the Sun has the rather high value of 5.

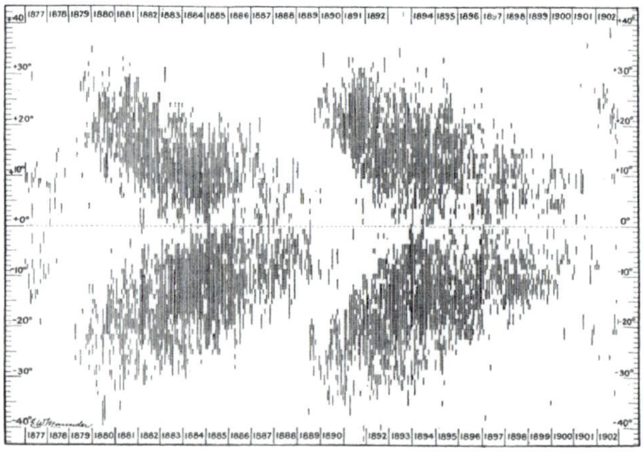

FIG. 8—DISTRIBUTION OF SPOT-CENTRES IN LATITUDE, ROTATION BY ROTATION, 1877–1902.

Figure: The butterfly diagram of sunspot statistics. Note in particular that the spots only exist very close to the equator; only a few spots appear at latitudes higher than (say) 30°. There is a well-established (but not perfect) symmetry of the phenomenon with respect to the equator

Only the antisymmetric (dipolar) parts of the solar magnetic field oscillate and reverses its sign during the 22-yrs cycle. The magnetic reversal happens at the poles during the maximum of the sunspot activity. As the latter is located rather close to the equator one can be sure that the solar activity phenomenon is a global phenomenon rather than a local one.

The Maunder Minimum

The amplitude of the cycle is far from being constant. The most prominent activity drop known was the Maunder minimum between 1670 and 1715. The latitudinal distribution of the few sunspots observed during the end of the Maunder minimum was highly asymmetric with respect to the equator. All spots except two or three in this period appeared on the southern hemisphere. The first cycle after the minimum, with maximum in 1706, existed almost exclusively in one hemisphere. There is empirical evidence that the magnetic cycles persisted through the solar Maunder minimum, as first found by Wittmann (1978, old sunspot data), Schröder (1992, auroral activity) and more recently by Beer.

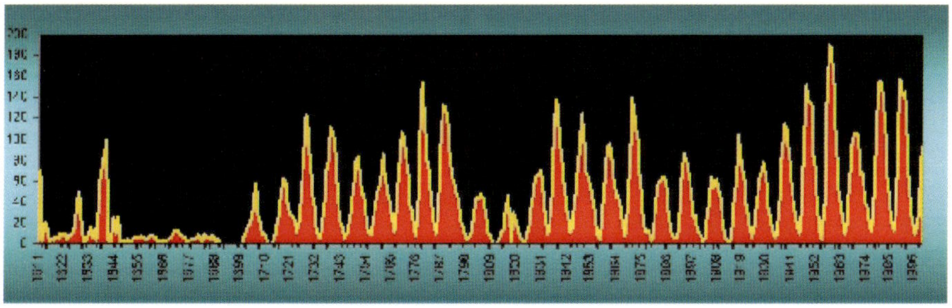

Figure: The time series of the sunspot numbers indicate a highly nonlinear behavior. Both cycle amplitude and length are obvious functions of time. From 1670--1715 a grand minimum ('Maunder minimum') is seen

Charbonneau (2001) reports an 'odd-even effect' of the cycle amplitudes. The odd-numbered cycles are stronger than their neighbouring cycles or -- which is the same -- the sunspot number of the complete 22-year cycle does not fluctuate too much from cycle to cycle.

Cycle Length Variations

Measurements of ^{14}C abundances in sediments and long-lived trees provide much longer time series than sunspot datasets. Vos found a secular periodicity of 80-90 yrs as well as a long-duration period of about 210 yrs. The measurements of atmospheric ^{14}C abundances suggested a periodicity of 2400 yrs which is also associated with a long-term variation of solar activity. The variety of frequencies found in solar activity may even indicate chaotic behavior as discussed by Knobloch & Landsberg (1996) on the basis of nonlinear mode-mixing models.

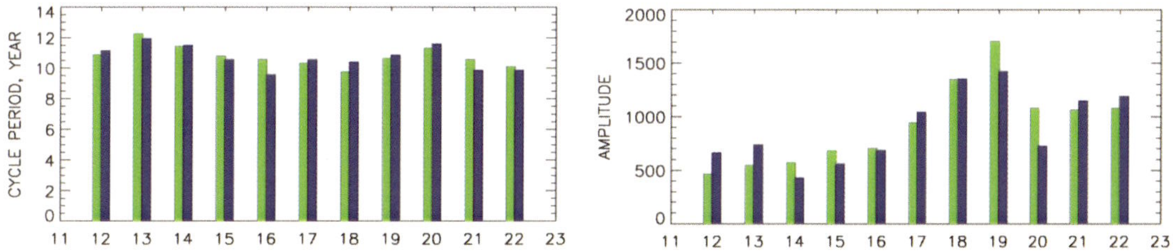

Figure: Cycle length (*left*) and cycle amplitudes (*right*) for 11 cycles separated for northern (green) and southern (blue) hemispheres

Another question concerns a possible systematic north-south asymmetry of the cycle length. If it is true that the grand minima result from simultaneous nonlinear oscillation of modes with equatorial symmetry and antisymmetry then systematic differences of the cycle length in the two hemispheres must appear. Both cycle lengths and amplitudes in the dataset of Hathaway are given in figure. There seems to be a slight anticorrelation between cycle length and cycle amplitude in the data: short cycles are followed by stronger cycles.

Any single cycle is asymmetric in time. The strong cycles are faster to reach their activity maximum than the weak cycles. This 'Waldmeier rule' can be used to produce successful predictions of the strength and duration of future activity cycles.

Solar-stellar Relations

The activity cycle of the Sun is not exceptional among the stars. The observation of chromospheric Ca-emission of solar-type stars yields activity periods between 3 and 30 yrs. Up to 15% of the solar-type stars, however, do not show any significant activity. This suggests that even the existence of the grand minima is a typical property of cool main sequence stars like the Sun. From ROSAT X-ray data Hempelmann find that up to 70% of the stars with a constant level of activity exhibit a rather low level of coronal X-ray emission.

The dependence of the cycle frequency on the rotation rate Ω is rather weak. Saar & Brandenburg (1999) report for the relation

$$\omega cyc \propto \Omega n (1)$$

the very small value $n \simeq -0.09$ for their full data-set. They derive values of 1.15, 0.8 and 0.4 for their groupings 'inactive', 'active' and 'superactive'. With the same data base Böhm-Vitense (2007) finds similar relations with $n \simeq 1$. The active stars are given in pink while the less active stars are given in blue and black. The latter really form an own branch with n of order unity while there is no clear rule for the fast-rotating stars. Considering only the shortest cycle for a given rotation rate, one finds the small value $n = 0.18$. For fast rotating stars the angular velocity itself seems to be not the only determinant of the cycle time.

Solar Dynamo

The fluid flows in the sun dives a magnetic dynamo. This physical process which produces the magnetic field of sun is known as the solar dynamo. It is possible to measure and calibrate, how much toroidal and radial magnetic field can grow and maintain. This solar magnetic field is not similar as the magnetic field of Earth.

The magnetic field is not fixed in two poles of the Sun. The magnetic field is continuously changing with time. The region with the high magnetic fields are called the sunspots. Since the magnetic field is varying constantly, the number sunspots are also varying in time. Over a solar cycle the number will increase and then decrease.

The solar dynamo is believed to be located at the overshoot zone or the convection zone. The plasma in the convection zone contains continuously moving electrically charged particles. These moving particles produce an electrical current. This generates magnetic fields and this field in turn generates electric current and this process continues as a loop. This will act as a solar dynamo.

For the magnetic field to act as a dynamo, the plasma flow should be turbulent, helical and fully three dimensional. Also, a differential rotation is necessary for the solar dynamo to work. In this, the continuous back and forth motion of plasma alters the azimuthal magnetic field to meridional magnetic field and vice versa.

In azimuthal magnetic field, the magnetic field points towards east-west and vice versa rather than north south. This loop of magnetic fields will continue until it is not get interrupted.

Theory

Schwabe's (1844) 11-year sunspot cycle, Carrington's (1863) differential rotation (of the solar surface) and Spörer's law (of the equatorward migration of sunspots during the cycle, 1894) are the basic properties of the solar activity that must be explained by the dynamo theory. As these three basics are understandable even with a *linear theory* (which already demonstrates the complexity of the problem) only a linear approach is here presented. There are many more nonlinear extensions of the models but the main principles and problems already appear in the given linear approach.

Convection Zone Dynamo

In the solar surface there is turbulence with velocity fluctuations of 1 km/s and correlation times of about 10 min. The magnetic diffusivity η_T formed by such high-speed short-living granular motion is of order 10^{12} cm^2/s. This value allows a skin depth of an electromagnetic oscillator with a timescale of 10 yrs of more than 100 000 km -- close to the depth of the solar convection zone (200 000 km). This is the background for the common believe that the entire solar convection zone should be the site of the solar dynamo. If this is true then also the large ratio of the mean cycle period and the rotation time of the stars becomes understandable. The basic time scale in convection zones is the correlation time of the turbulent cells which at the bottom of the solar convection zone reaches the value of the rotation time (25 days). In a thick convection shell the number $\tau cyc/\tau corr$ reflects the square of the ratio of the stellar radius to the correlation length so that numbers exceeding 100 are easily possible.

Most solar dynamo models also operate in a linear way. Rather weak seed fields are amplified in the sense of a self-excitation of large-scale magnetic fields. The global magnetic field can be described by a superposition of a poloidal field (the field projected in a meridional plane) and the toroidal (azimuthal) field component. Both field components would independently decay if there are no extra processes which maintain the magnetic field against the Ohmic decay which would destroy the fields after about 10 yrs.

A dynamo can thus only operate if processes exist which transform poloidal fields into toroidal fields and v.v. One of the processes is well-known: if the angular velocity in the convection zone is not uniform, then it easily transforms poloidal fields into toroidal fields by wiggling them up. The effectivity of this transformation scales with the magnetic Reynolds number of the differential rotation, i.e $B\phi/Br \simeq \delta\Omega\, D^2/\eta_T$ (with D as the depth of the convection zone) which is about 100 for $\eta_T = 5 \cdot 10^{12}$ cm²/s and 5000 for $\eta_T = 10^{11}$ cm²/s . It is thus easy to understand that the amplitude of the toroidal field of the Sun exceeds the amplitude of the poloidal field. Obviously, the observed factor of about 1000 needs a rather small eddy diffusivity.

Electromotive Force (EMF) in Turbulent Plasma

More complicated is the regeneration of the poloidal field from the toroidal one. In principle it can be done by the action of a meridional flow but in this case the resulting magnetic field must be nonaxisymmetric in a large scale (Cowling theorem). One has thus to look for the hydromagnetic properties of the convection/turbulence in the solar convection zone. It is in particular the structure of the turbulent electromotive force $E = \langle u' \times B' \rangle$ which might be important. It appears in the Ohm's law for mean-field quantities

$$\langle J \rangle = \sigma(\langle E \rangle + \langle u \rangle \times \langle B \rangle + E)$$

and plays the central role in the mean-field dynamo theory.

The simplest electrodynamic effect of a turbulent field of electrically conducting matter is its diamagnetism, i.e. the tendency to transport a mean magnetic field into regions of lower turbulence intensity. It can simply be written as

$$E = \gamma \times \langle B \rangle + \ldots$$

with $\gamma = -\frac{1}{2}\nabla\eta_T$. The field is thus transported towards the minimum of the eddy diffusivity η_T. The effect is also known as turbulent pumping; it is the reason that in dynamo simulations the flux is mainly transported downwards rather than upwards as the various buoyancy effects would suggest. If the downward pumping is too strong then the dynamo does not operate. For $\eta_T \simeq 1012$ cm2/s the amplitude of the downward pumping exceeds the value of 1 m/s.

Another turbulence-induced term runs with the mean-field electric current, i.e.

$$E = \cdots - \eta_T \mathrm{curl}\langle B \rangle,$$

which describes the strongly enhanced dissipation of the field by the turbulence. The ratio of the eddy diffusivity η_T to the microscopic diffusivity of the gas is extremely high. If the basic stellar rotation is included into the computations of the EMF the magnetic eddy diffusivity forms a complicated tensor with many components and consequences also for the dynamo theory.

The Alpha-effect Due to Rotation

The transformation of the toroidal field to the poloidal field needs another term. It is the alpha-effect by Steenbeck, Krause & Rädler (1966) which is the basis of almost all of the solar dynamo models developed so far. After the alpha-effect concept in rotating turbulences of conducting matter a relation

$$E = \alpha\, \langle B \rangle + \ldots$$

exists so that by the action of the turbulence an electric current flows along the magnetic field lines.

The current along the toroidal field is always accompanied by a poloidal field, hence the loop can be closed and the magnetic field can be maintained against its Ohmic decay.

An alpha-effect only exists if the star rotates. Additionally there *must be a stratification of density and/or turbulence intensity* otherwise the alpha-effect is zero. The following argument describes the situation. The quantity alpha is a pseudo-scalar as it connects the polar vector E with the axial vector $\langle B \rangle$. The simplest pseudoscalar one can form in turbulent flows is $\langle u' \cdot \mathrm{curl} u' \rangle$ that describes a helicity in the turbulence field. Helicity means that the turbulence is not mirror-symmetric, i.e. (say) left-handed helical motions are more frequent than right-handed ones or v.v. This is only possible if the star rotates.

The simplest pseudoscalar one can construct with *global quantities* is the scalar product of the vector of the density stratification $g=\nabla \log\rho$ and the vector of the angular velocity of the basic rotation so that $\alpha \propto g \cdot \Omega$ should be true. The sign of this quantity differs in the northern and the southern hemisphere and so both the alpha-effect and also the helicity do. At the equator the alpha-effect vanishes, but it exists at the poles. For mesogranulation the helicity can directly be observed at the solar surface. It proves to be negative (positive) at the northern (southern) hemisphere.

There is indeed a theoretical relation of the alpha-effect and the numerical value of the helicity, i.e.

$$\alpha \simeq -\tau_{\mathrm{corr}} \langle u' \mathrm{curl} u' \rangle, \qquad (6)$$

which has been confirmed by many numerical simulations. Highly nonlinear calculations with forced turbulence also reveal the contributions of *magnetic helicity* to the EMF which can also be observed at the solar surface.

Dynamo Waves

The simplest dynamo models even work in low dimensions producing by the common action of alpha-effect and differential rotation traveling magnetic waves with a frequency of

$$\omega_{\mathrm{cyc}} \simeq \left(\frac{\alpha^2}{\eta} \left(r \frac{\mathrm{d}\Omega}{\mathrm{d}r} \right)^2 \right)^{\frac{1}{3}}$$

To obtain a cycle time of about 20 yrs an alpha of order 1 m/s is thus necessary. Simulations of the alpha-effect in the solar convection zone yield $\alpha \simeq 10-(1...2)\ u_T$ with u_T as the characteristic turbulence velocity so that with $u_T \simeq 100$ m/s indeed an alpha-effect of 1 m/s results. The expression also explains the rather weak dependence of the cycle times of stellar dynamos on the rotation time. Theory and observation of the stellar differential rotation lead to the common result that the rotation law in stars hardly depends on their rotation rate.

Also the alpha-effect does not depend strongly on the rotation rate. Ossendrijver report an increase of the alpha-effect (in horizontal direction) by a factor of only 2.5 if the rotation rate varies by two orders of magnitude. Hence, the result complies with the observed small exponent *n*.

The Differential Solar Rotation

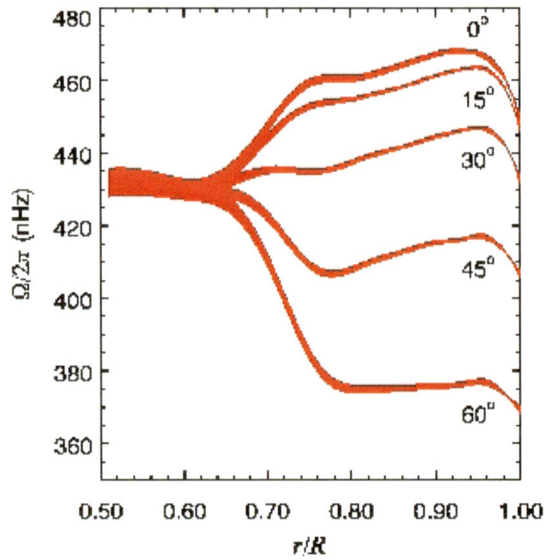

Figure: The internal rotation law of the Sun from the inversion of SOHO data.
The curves are marked with the latitude

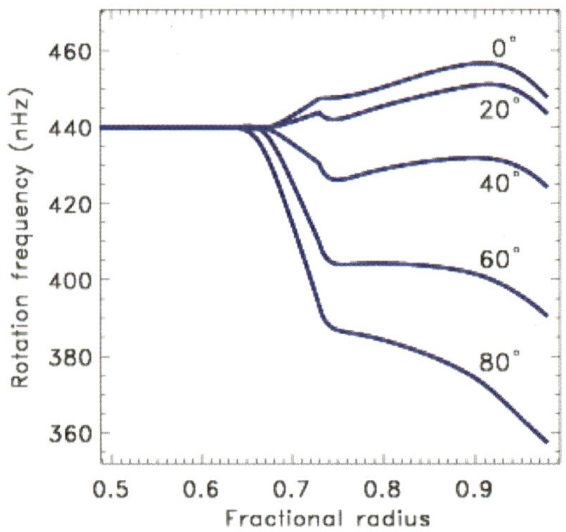

Figure: The solar rotation law computed for 25 days rotation period . The model includes two parameters only for the anisotropic correlation tensor of the turbulence in the convection zone

The rotation law within the solar convection zone is known by the inversion procedures of the helioseismology. Formally, the observed equatorial acceleration at the surface proves to be the result of a strong polar subrotation ($\partial\Omega/\partial r<0$) and a weaker equatorial superrotation ($\partial\Omega/\partial r>0$) inside the convection zone. The *strong polar subrotation forms the main problem for the solar dynamo theory* as naturally the strongest fields are there induced which, however, are not observed. The solar core rotates rigidly with approximately the same angular velocity as the surface at midlatitudes. The solution of the Reynolds equation on the basis of the (also turbulence-induced) Lambda-effect for rotating density-stratified turbulence do indeed explain these observations.

The stellar-surface rotation laws which are empirically known so far lead to the result that the absolute equator-pole differences of the solar-type stars hardly differ from star to star.

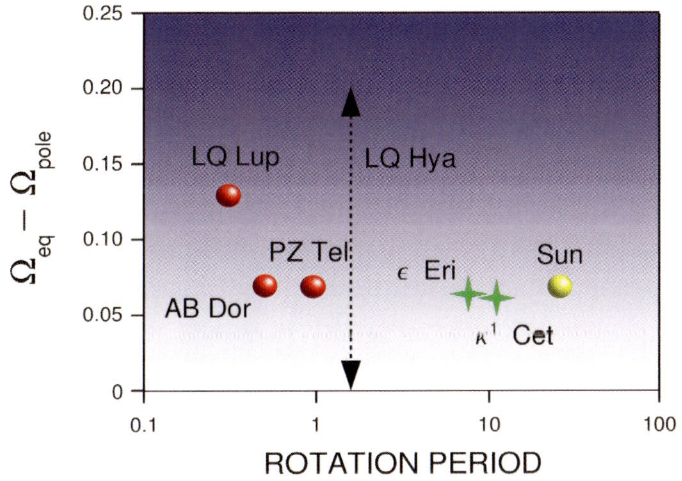

Figure: The sofar known equator-pole differences (in rad/day) of very young stars, the MOST observations and the present-day Sun. For LQ Hya the equatorial acceleration is reported as highly time-dependent. LQ Lup is too fast for present-day theory. Rotation period in days

Data from the MOST satellite recently led to the detection of rotation laws for two new young solar-type stars with rotation periods of about 10 days. For ϵ Eri it is $\delta\Omega$=0.061 rad/day and for κ^1Ceti it is $\delta\Omega$=0.064 rad/day. Note the almost perfect coincidence of these values with the solar value. The two new examples are of extraordinary meaning insofar as their rotational rates start to fill the gap between the fast rotating very young stars and our (old) present-day Sun. A characteristic value of the equator-pole difference of Ω for stars (including the Sun) is $\delta\Omega$=0.07 rad/day . This value excellently complies with the results of the theory of the rotation laws of stars with outer convection zones.

Meridional Flow

All nonuniform rotation laws exist together with meridional flows as the centrifugal force of nonrigid rotation cannot longer be written as a conservative force. The flow which occurs in the simulations at the bottom of the convection zone drifts always equatorwards with an amplitude of um$\simeq 5^{m}/s$. Such a flow needs slightly more than 7 yrs to travel from the pole to the equator. However, with η_T=1012 cm²/s the magnetic Reynolds number Rm=$u^{m}R/\eta_T$ (R ~ 700.000 km, solar radius) does not strongly exceed unity so that the meridional flow will not be important in the dynamo equation. Only if the magnetic eddy diffusivity is reduced by (say) a factor of 10 the meridional circulation dramatically changes the dynamo regime by fixing the cycle time and transporting the azimuthal magnetic field towards the equator.

Numerical Models

The dynamo equation

$$\frac{\partial\langle B\rangle}{\partial t} = \mathrm{curl}\big((u+\gamma)\times\langle B\rangle + \alpha\langle B\rangle - \eta_T \mathrm{curl}\langle B\rangle\big)$$

(with the pumping term included) must be solved with u as the given large-scale flow pattern in the solar convection zone. Simple models only apply the solar rotation law . If the fields are weak enough then any magnetic feedback to the terms of the electromotive force can be ignored. Calculations with this assumption are called as 'kinematic' dynamo models. Nonlinear dynamo models are much more complicated.

Distributed Dynamos

Let the magnetic eddy diffusivity η_T be homogeneous within the entire convection zone (thickness $D \simeq 200.000$ km) while the solar core is considered as a perfect conductor. The alpha-effect always runs with $\cos\theta$ so that it peaks at the poles. If the angular velocity has only a radial gradient, then the cycle time turns out to be of order $\tau cyc \simeq 0.26 DR/\eta_T 0$. The value $\eta T \simeq 10^{12}$ cm^2/s is thus the reference value leading to a cycle time of 10 yrs.

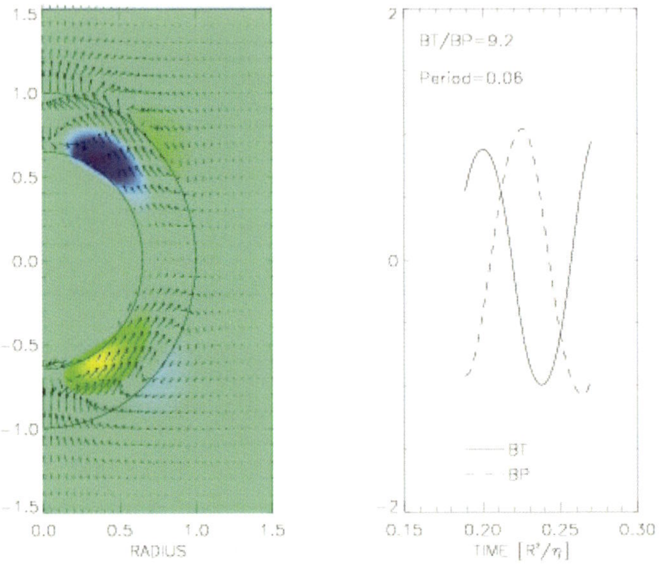

A solar dynamo model with alpha-effect in the entire convection zone and ηT=5·1012 cm2/s

Left: The geometry of the induced magnetic field. The colors represent the magnitude of the toroidal field while the arrows form the field lines of the poloidal field. Note the equatorial asymmetry. Right: Phase relation of the field components (BT toroidal, BP poloidal). The cycle time is given in units of the diffusion time (here 33 yrs).

With the observed solar rotation law and positive (in the northern hemisphere) depth-independent alpha-effect Eq. yields magnetic fields which are represented in figure. One finds a radial drift of the toroidal field belts but not an equatorial migration, i.e. the butterfly diagram figure above can not be explained. The parity of the solution is dipolar, the toroidal magnetic amplitudes are much too weak. Also the cycle time is too short.

The model can be improved if the alpha-effect is concentrated to a thin layer at the bottom of the convection zone. Then the amplitude ratio exceeds values of 200 but one always gets more magnetic belts than needed. The butterfly diagram appears at too high latitudes and will never reach the low latitudes which are necessary to reflect the solar observations correctly.

Too simple kinematic $\alpha\Omega$ -dynamos working in the convection zone with positive alpha-effect and the solar rotation law are thus not able to fulfill the main observational constraints.

Boundary-layer Dynamos

The idea of the boundary-layer (BL) dynamo bases on the findings that i) the alpha-effect in the bottom of the solar convection zone, where the convection overshoots into the stably stratified radiative zone, becomes negative and ii) the radial gradients of the angular velocity peak there. One could thus indeed realize an equatorwards drift of the toroidal field belts in case the alpha-effect vanishes at the poles. If the boundary layer, however, is too thin then the cycle time becomes too short for $\eta_T = 10^{12}$ cm²/s and also the number of magnetic belts becomes too high. Indeed, after current convection zone simulations the overshoot phenomenon restricts to a layer with less than 10.000 km depth. It is hard to believe that in such a thin layer the averaging procedure for the turbulence can be done leading to a mean-field formulation of the EMF. Current dynamo models thus prefer to work with the positive alpha-effect dominating the entire convection zone. In these cases the butterfly diagram can only result from the action of the meridional circulation flowing towards the equator at a radius where the toroidal magnetic field has its maximum.

Advection-dominated Dynamos

The inclusion of the meridional flow has strong impacts to the mean-field dynamo models only when the eddy diffusivity is so low that the magnetic Reynolds number of the flow reaches values of the order of 10^3. The consequences for the dynamo regime are drastic. Models with rather small values of the eddy diffusivity have been a subject of intense numerical investigation. Models with a positive alpha-effect and high magnetic Reynolds number provide correct cycle periods, equatorial parity and produce rather good butterfly diagrams.

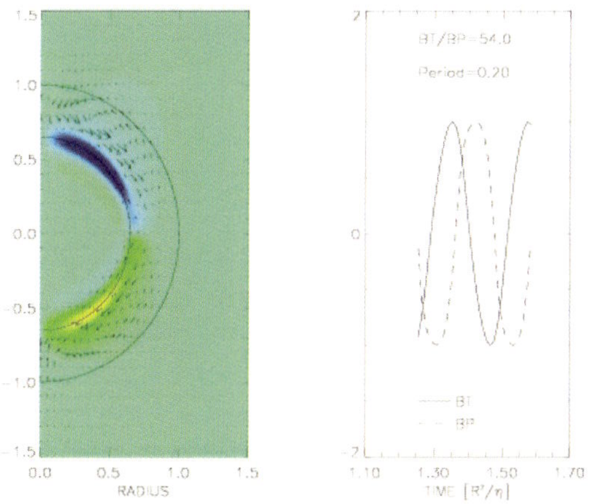

Figure: Advection-dominated solar dynamo with alpha-effect located at the bottom of the convection zone where the magnetic diffusivity has been assumed as very small ($2*10^{10}$ cm²/s) and with a flow amplitude of 5 m/s

In the simplest model the alpha-effect exists in the whole convection zone and the drift amplitude lies between 2 m/s and 10 m/s. Almost always solutions with the (wrong) quadrupole solution occur for the lowest eigenvalue. Because of the small value of the eddy diffusivity the cycle time becomes too long compared with the 22 yrs of the Sun. The toroidal field belts are migrating equatorwards but the maximum field amplitude still exists in the polar region.

These results can drastically be improved if the alpha-effect is located mainly at the bottom of the convection zone. The η_T is 10^{12} cm²/s in the bulk of the convection zone but it is smaller (We choose the more complex situation the rotation is also unstable without the density stratification ($2*10^{10}$

cm^2/s) at its bottom. Now the dipolar parity is indeed preferred. The meridional flow advects the field equatorwards producing a butterfly diagram of the observed type. For such models one finds that the inclusion of the meridional circulation into models with low values of the magnetic diffusivity

- improves the cycle time problem
- transports the fields towards the equator
- favors the dipolar solutions against the quadrupolar solutions
- produces strong enough toroidal fields
- avoids the high number of field belts which often appears in thin layer solutions.

The needed circulation seems really to exist. The surface counterpart of this flow moving towards the poles with an amplitude of about 15 m/s has been observed. However, the reason for the small value of the eddy diffusivity (i.e. the high magnetic Prandtl number) is still unknown. The dependence of the cycle time on the amplitude of the meridional flow at the bottom of the convection zone is shown in figure as a summary of various model calculations. One can easily see that the cycle time is much too large for too slow circulation but for drift rates in excess of 6 m/s the resulting cycle times well reproduce the observed cycle time of 22 yrs.

Simulation Models

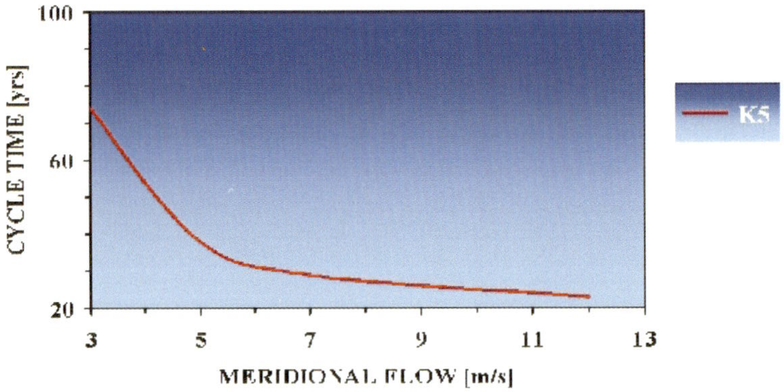

Figure: The cycle times for the dipolar modes of advection-dominated dynamo models of Bonanno. There is a clear anticorrelation of the drift velocity and the cycle time. Note the very long cycle times for too slow meridional circulation which are due to the small values of the used turbulent magnetic diffusivity

Many of the known mean-field phenomena can also be simulated with high-developed supercomputer codes which numerically solve the first-principle equations of magnetohydrodynamics. By heating a convectively unstable spherical shell from below rotating magnetoconvection results which transports angular momentum and generates large-scale magnetic fields. Basic details of the mean-field calculations are reproduced. The toroidal magnetic fields in the convection zone appears to be weak and highly nonaxisymmetric but the field in the 'tachocline' layer below the bottom of the convection zone becomes strong (~ 3000 G), axisymmetric and antisymmetric with respect to the equator. The latter result can be described as due to the action of an alpha-effect in the convection zone. Also the rotation law in the convection zone shows the main properties of the observed pattern. Still some of the results are consequences of special inputs but basic ingredients of the solar dynamo model -- such as the alpha-effect in the convection zone, differential rotation, turbulent pumping and effective toroidal stretching of the magnetic field lines -- can be obtained by direct numerical simulations. However, as the solar cycle lasts more than 200 rotation times it will be a highly innovative work to reproduce a full solar cycle with the supercomputer.

References

- These times come from: Byrne, J. Neutrons, Nuclei, and Matter, Dover Publications, Mineola, New York, 2011, ISBN 0486482383, p 8
- Sun-core-rotates-faster-surface: phys.org, Retrieved 08 June 2018
- García, Ra; Turck-Chièze, S; Jiménez-Reyes, Sj; Ballot, J; et al. (Jun 2007). "Tracking solar gravity modes: the dynamics of the solar core". Science. 316 (5831): 1591–3. Bibcode:2007Sci...316.1591G. doi:10.1126/science.1140598. ISSN 0036-8075. PMID 17478682
- Convection-zone-the-journey-continues: planetfacts.org, Retrieved 28 May 2018
- Karl S. Kruszelnicki (17 April 2012). "Dr Karl's Great Moments In Science: Lazy Sun is less energetic than compost". Australian Broadcasting Corporation. Retrieved 25 February 2014
- Sun-photosphere: scied.ucar.edu, Retrieved 18 May 2018
- Solar-chromosphere: scied.ucar.edu, Retrieved 15 May 2018
- Pascale Ehrenfreund; et al., eds. (2004). Astrobiology: future perspectives. Dordrecht [u.a.]: Kluwer Academic. ISBN 1402023049. Retrieved 28 August 2014
- Suns-photosphere-5218439: sciencing.com, Retrieved 30 June 2018

5

The Solar System

The Solar System is a gravitationally bound system comprising of the Sun and the planets. Astrophysics studies the solar system, with an emphasis on its formation and evolution, ring system, planetary system, etc. which have been discussed in this chapter.

The Solar system (or solar system) is the home stellar system for human beings and all known forms of life. The solar system comprises the Sun, all the objects gravitationally bound to it, and the heliosphere, an enormous magnetic bubble enclosing most of the known solar system, including the solar wind and the entire solar magnetic field. Objects bound gravitationally to the Sun include eight planets, their 165 known moons, at least three dwarf planets (including Pluto) with their four known moons, and billions of small bodies, classified primarily as asteroids, meteoroids, or comets.

In broad terms, the solar system comprises regions starting with the Sun and extending outward from there successively to:

- Four terrestrial inner planets
- The asteroid belt of small rocky bodies
- Four gas giant outer planets
- The Kuiper belt of icy objects
- The scattered disc
- The heliopause
- The hypothetical Oort cloud

The solar system major bodies drawn to scale

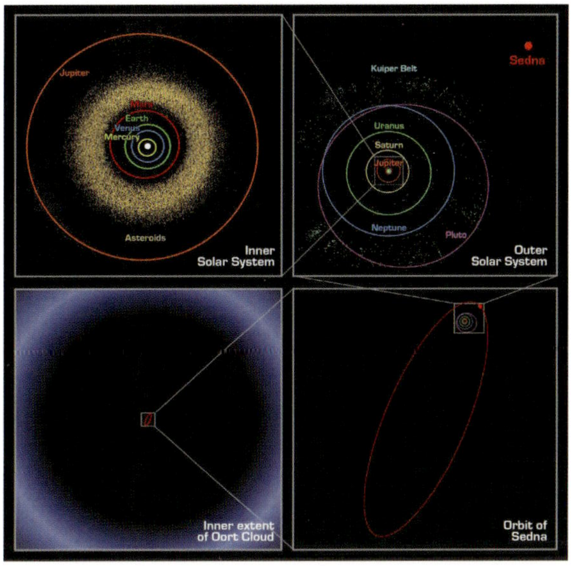

The solar system major orbits drawn to scale (clockwise from top left)

All solar system objects are immersed in either the solar plasma, known as the solar wind, the interstellar plasma of the Milky Way galaxy, or a mixture of the two in a region lying just inside the heliopause, the diaphonous boundary of the heliosphere.

The eight planets, in order of their distances from the Sun, are: Mercury, Venus, Earth, Mars, Jupiter, Saturn, Uranus, and Neptune. Six of the eight planets are in turn orbited by natural satellites, usually termed “moons” after Earth’s Moon, and each of the outer planets is encircled by planetary rings of dust and other particles. All of the planets, including the earth are surrounded by a teardrop shaped magnetosphere generated by the interaction of the magnetic field of the solar wind with that of each planet. All the planets except Earth are named after gods and goddesses from Greco-Roman mythology.

The solar system’s position in the Milky Way galaxy and the Earth’s position in the solar system are remarkably well-suited for the emergence and sustenance of life on Earth. The solar system occupies the Milky Way galaxy’s relatively narrow “habitable zone” where the intensity of gamma rays, x-rays, and other radiation from surrounding stars is not too great to block the development of life by sterilizing planetary surfaces. This zone is comfortably removed from the intense concentration of stars closer to the galactic centre, but not too close to the outer margins of the galaxy where there are not enough of the heavier elements to support formation of planets. Within the solar system, the Earth is neither so close to the sun as to be too hot for life nor so far away as to be too cold. Had the Earth been much closer to the Sun or much farther away, life would not have arisen on the planet.

Historically, the solar system has been the key performer in the greatest and grandest show on earth, the drama of the sky with the sun moving across it each day and with the fixed stars of the constellations providing each night a slowly moving backdrop against which the more agile moon, and five visible planets Mercury, Venus, Mars, Jupiter, and Saturn performed their routines. Massive monuments such as Stonehenge in England and pyramids in Egypt and the Americas attest to ancient peoples’ profound efforts to observe and predict movements of the solar bodies, and the active fields today of both astrology and solar astronomy demonstrate the continued importance of the solar system in human life. In western cultures, the names of the days of the week are derived from names of the planets or the names of gods associated with them. The current solar system model with the powerful central sun guiding from afar the planets and other solid bodies and radiating the light and heat that sustain life itself embodies an archetype that has been adopted into other fields.

Major features of the solar system (not to scale): The Sun, the eight planets, the asteroid belt containing the dwarf planet Ceres, the dwarf planet Pluto (in outermost orbit), and a comet (the dwarf planet Eris is not shown).

The principal component of the solar system is the Sun (astronomical symbol ⊙); a main sequence G2 star that contains 99.86 percent of the system's known mass and dominates it gravitationally. Because of its large mass, the Sun has an interior density high enough to sustain nuclear fusion, releasing enormous amounts of energy, most of which is radiated into space in the form of electromagnetic radiation, including visible light. Jupiter and Saturn, the two largest bodies orbiting the Sun, account for more than 90 percent of the system's remaining mass (the postulated Oort cloud, should its existence be confirmed, would also hold a substantial percentage).

The ecliptic viewed in sunlight from behind the Moon in this image taken by the Clementine spacecraft. From left to right: Mercury, Mars, Saturn

Most objects in orbit around the Sun lie within the same shallow plane, called the *ecliptic*, which is roughly parallel to the Sun's equator. The planets lie very close to the ecliptic, while comets and

Kuiper belt objects often lie at significant angles to it. All of the planets and most other objects orbit with the Sun's rotation in a counterclockwise direction as viewed from a point above the Sun's north pole. There is a direct relationship between how far away a planet is from the Sun and how quickly it orbits (Kepler's laws of planetary motion). Mercury, with the smallest orbital circumference, travels the fastest, while Neptune, being much farther from the Sun, travels slower.

A planet's distance from the Sun varies in the course of its year. Its closest approach to the Sun is known as its *perihelion*, while its farthest point from the Sun is called its *aphelion*. Though planets follow nearly circular orbits, with perihelions roughly equal to their aphelions; many comets, asteroids and objects of the Kuiper belt follow highly elliptical orbits, with large differences between perihelion and aphelion.

Astronomers most often measure distances within the solar system in astronomical units (AU). One AU is the average distance between the Earth and the Sun, or roughly 149,598,000 kilometers (93 million miles). Pluto is roughly 38 AU from the Sun, while Jupiter lies at roughly 5.2 AU.

Representing the planets and the sun as a unit solar system in a single graphic presents major challenges because both the individual bodies (diameters ranging from 5,000 to 1,390,000 kilometers) and the planets' distances from the sun (average orbital radii ranging from .4 to 38 AU) are so greatly varied. As a result, many representations of the solar system distort the planetary orbits, with equal spacing between each object. In actuality, the general pattern is that the farther a planet or a belt is from the Sun, the greater the distance between it and the previous orbit. Attempts have been made to find a correlation between these distances (an early one was Bode's Law), but to date there is no accepted theory that explains the respective orbital distances.

Classification of Solar System Bodies

A planet, as defined by the IAU General Assembly on August 24, 2006, is any body in orbit around the Sun that (a) has enough mass to form itself into a near-spherical shape, and (b) has cleared or captured its immediate neighbourhood of all smaller objects. Eight objects in the solar system currently meet these criteria: Mercury, Venus, Earth, Mars, Jupiter, Saturn, Uranus, and Neptune.

The IAU also used the term dwarf planet, a newly defined class of solar system bodies. The key difference between planets and dwarf planets is that while both are required to orbit the Sun and be of large enough mass that their own gravity pulls them into a nearly round shape, dwarf planets are not required to "clear their neighbourhood" of other celestial bodies. Three objects in the solar system are currently included in this category. They are: Pluto, the largest known Kuiper belt object; Ceres, the largest object in the asteroid belt; and the scattered disc object Eris. The IAU is evaluating other known objects to see if they fit within the definition of dwarf planets. The most likely candidates are some of the larger asteroids and several trans-Neptunian objects.

The remaining objects in the solar system were classified as small solar system bodies (SSSB). The IAU used the term SSSB to describe solar system objects that are neither planets nor dwarf planets.

> *All other objects orbiting the Sun shall be referred to collectively as "Small Solar System Bodies" These currently include most of the Solar System asteroids, most Trans-Neptunian Objects (TNOs), comets, and other small bodies.*

As of 2006, the IAU considers the following bodies to be SSSBs:

1. all asteroids except Ceres
2. all centaurs
3. all trans-Neptunian objects, including Kuiper belt and scattered disc objects, with the exception of Pluto and Eris
4. all comets

Sun

The Sun as seen from Earth

The Sun is the solar system's parent star, and far and away its chief component. It is classified as a moderately large yellow dwarf. This name, however, is misleading, because on the scale of stars in our galaxy, the Sun is rather large and bright. Stars are classified based on their position on the Hertzsprung-Russell diagram, a graph that plots the brightness of stars against their surface temperature. Generally speaking, the hotter a star is, the brighter it is. Stars that follow this pattern are said to be on the "main sequence," and the Sun lies right in the middle of it. This has led many astronomy textbooks to label the Sun as "average," although stars brighter and hotter than it are rare, whereas dimmer and cooler stars are common. The vast majority of stars are dim red dwarfs, though they are under-represented in star catalogs because we can observe only those that are relatively close to the Sun.

According to current theories of stellar evolution, the Sun's position on the main sequence means that it is in the "prime of life" for a star. In other words, it has not yet exhausted its store of hydrogen for nuclear fusion, and thus it has not been forced (as older red giants must) to fuse more inefficient elements such as helium and carbon. As it ages, the Sun is growing increasingly bright. Early in its history, it was roughly 75 percent as bright as it is today. Calculations of the ratios of hydrogen and helium within the Sun suggest it is roughly halfway through its life cycle and will eventually begin moving off the main sequence becoming larger, brighter, and redder, and becoming a red giant in about five billion years.

The Sun is a "population I" star meaning that it is fairly new in galactic terms, having been born in the later stages of the universe's evolution. As such, it contains far more elements heavier than hydrogen and helium (metals, in astronomical parlance) than older, population II stars, such as those found in globular star clusters. Because elements heavier than hydrogen and helium were formed in the cores of ancient and exploding stars, the first generation of stars had to die before the universe could be enriched with those elements. For this reason, the very oldest stars contain very little "metal," while stars born later have more. This high "metallicity" is thought to have been crucial for the Sun to develop a planetary system, because planets form from the accretion of metals.

The Sun ejects the solar wind, a pulsing stream of electrons, ions, neutral atoms, and magnetic fields, at speeds greater than 2 million kilometers per hour. It thus creates a very tenuous "atmosphere" (the heliosphere), that permeates the solar system for at least 100 AU. This environment is known as the interplanetary medium. Small quantities of cosmic dust (some of it arguably interstellar in origin) are also present in the interplanetary medium and are responsible for the phenomenon known as zodiacal light. The influence of the Sun's rotating magnetic field on the interplanetary medium creates one of the largest structures in the solar system: the heliospheric current sheet.

Inner Planets

The inner planets, from left to right: Mercury, Venus, Earth, and Mars (sizes to scale)

The four inner (or terrestrial) planets are characterized by their dense, rocky composition, few or no moons, and lack of ring systems. They are composed largely of minerals with high melting points, such as silicates, that form their solid crusts and semi-liquid mantles. In addition, metallic dust grains such as iron form the cores of these planets. Three of the four inner planets have atmospheres. All have impact craters, and all but one possess tectonic surface features, such as rift valleys and volcanoes. The term *inner planet* should not be confused with *inferior planet*, which designates planets that are closer to the Sun than the Earth is (namely, Mercury and Venus).

Mercury

Mercury (0.4 AU), the closest planet to the Sun, is also the least massive of the planets, at only 0.055 of Earth's mass. Mercury has a very thin atmosphere, consisting of atoms blasted off its surface by the solar wind. Mercury is so hot that these atoms quickly escape into space. Thus, in contrast to the Earth and Venus, which have stable atmospheres, Mercury's atmosphere is constantly being replenished. Mercury is surrounded by extremely small amounts of helium, hydrogen, oxygen, and sodium. This envelope of gases is so thin that the greatest possible atmospheric pressure (force exerted by the weight of gases) on Mercury would be about 2×10^{-12} kg/cm^2 (3×10^{-11} psi). The atmospheric pressure on Earth is about 1.03 kg/cm^2 (14.7 psi). It has no natural satellites, and, to date, no observed geological activity other than that produced by impacts. Its relatively large iron core and thin mantle have not yet been adequately explained. Some have postulated that its outer layers may have been stripped off by a giant impact, and that it was prevented from fully accreting by the Sun's gravity. The MESSENGER spacecraft should aid in resolving this issue when it arrives in Mercury's orbit in 2011.

Venus

Venus (0.7 AU), the first truly terrestrial planet, is of mass comparable to that of the Earth (0.815 of Earth mass) and has been called "Earth's twin." Like Earth, it possesses a substantial atmosphere, a thick silicate mantle around an iron core, and evidence of one-time internal geological activity, such as volcanoes. It is, however, much drier than Earth, its atmosphere is 90 times as dense, and it is composed overwhelmingly of carbon dioxide and sulfuric acid. Evidence suggests that the crust of Venus, unlike Earth's crust, is not divided into tectonic plates, but rather comprises a single, thick rind. Venus has no natural satellite. Also, it is the hottest planet, despite being farther from the Sun than Mercury, with temperatures reaching more than 400 °C. This is most likely due to the large amount of greenhouse gases (gases that trap heat) in the atmosphere.

Earth

The largest and densest of the inner planets, Earth (1 AU) is the only planet known to have life. It is also the only one to demonstrate unequivocal evidence of current geological activity. Its liquid hydrosphere, unique among the terrestrial planets, is probably the reason Earth is also the only planet where multi-plate tectonics has been observed because water acts as a lubricant for subduction.

The Earth's atmosphere is radically different from that of the other terrestrial planets, having been altered by the presence of life to contain 21 percent free oxygen. Its satellite, the Moon, is unusually large compared with the size of the Earth, causing significant effects on Earth such as ocean tides. In addition, the relative sizes and distances of the Moon and Sun enable us to observe a total solar eclipse. Earth's magnetic field protects its atmosphere from interacting with the solar wind, except in the vicinity of the poles where a variable influx of charged particles from the solar wind at times generates the shimmering Northern and Southern Lights.

Mars

Mars (1.5 AU), at only 0.107 of Earth's mass, is less massive than either Earth or Venus. It possesses a tenuous atmosphere of carbon dioxide. Its surface, peppered with vast volcanoes and rift valleys such as Valles Marineris, shows that it was once geologically active and recent evidence suggests this may have continued until relatively recently. Mars possesses two tiny moons, Deimos and Phobos, thought to be captured asteroids. Evidences from Martian meteorites, spacecraft observations, and rover experiments suggest that Mars may have had liquid water in the past, and some think that Mars may have harbored some primitive forms of life as well.

Asteroid Belt

Asteroids are mostly small solar system bodies that are composed in significant part of rocky, nonvolatile minerals. The main asteroid belt occupies an orbit between Mars and Jupiter, between 2.3 and 3.3 AU from the Sun. It is thought to consist of pieces that, due to the gravitational interference of Jupiter, failed to coalesce into a small terrestrial planet.

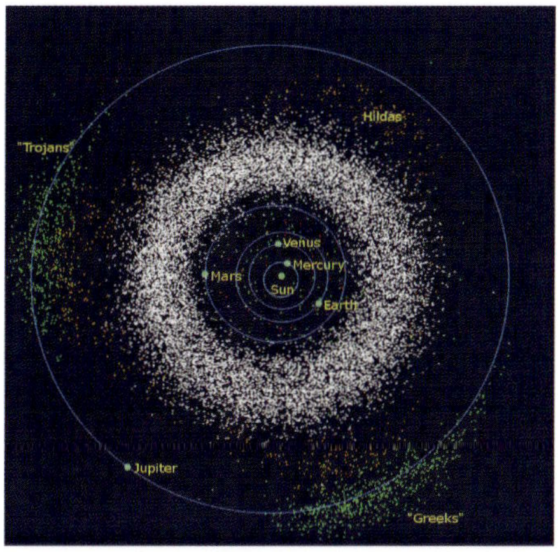

Image of the main asteroid belt and the Trojan asteroids

Asteroids range in size from hundreds of kilometers in diameter down to dust particles. All asteroids save the largest, Ceres, are classified as small solar system bodies. A number of other asteroids such as Vesta and Hygeia could potentially be reclassified as dwarf planets if it can be conclusively shown that they are spherical. The asteroid belt contains tens of thousands and potentially millions of objects over one kilometer in diameter. Despite their large numbers, the total mass of the main belt is unlikely to be more than a thousandth of the Earth's mass. In contrast to its various depictions in science fiction, the main belt is very sparsely populated spacecraft routinely pass through it without incident. Asteroids with a diameter of less than 50 meters are called meteoroids.

Ceres

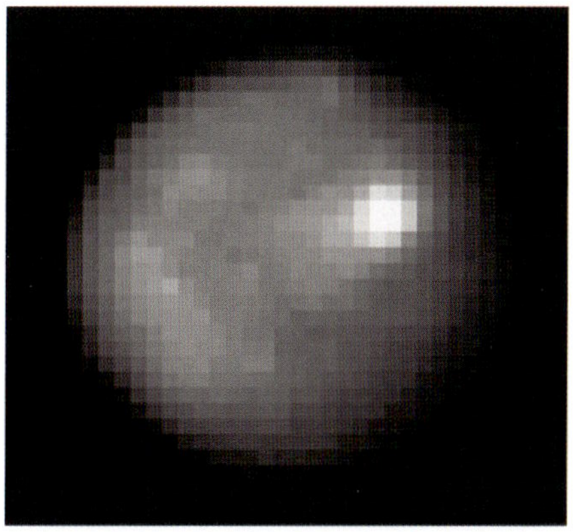

Ceres, as imaged by the Hubble Space Telescope

Ceres (2.77 AU) is the largest astronomical body in the asteroid belt and the only known dwarf planet in this region. It has a diameter of slightly under 1,000 kilometers large enough for its own gravity to pull it into a spherical shape. Ceres was considered a planet when it was discovered in the nineteenth century, but it was reclassified as an asteroid when further observations revealed additional asteroids. Its location coincides with what is predicted by Bode's Law.

Asteroid Groups

Asteroids in the main belt are subdivided into asteroid groups and families, based on their specific orbital characteristics. Asteroid moons are asteroids that orbit larger asteroids. They are not as clearly distinguished as planetary moons, sometimes being almost as large as their partners. The asteroid belt also contains main-belt comets, which may have been the source of Earth's water.

Trojan asteroids are located in either of Jupiter's "Lagrange points," which are gravitationally-stable regions leading and trailing the planet in its orbit. The term "Trojan asteroids" is also sometimes used for asteroids in the Lagrange points of other planets as well.

The inner solar system is also dusted with rogue asteroids, many of which cross the orbits of the inner planets.

Outer Planets

From top to bottom: Neptune, Uranus, Saturn, and Jupiter (sizes not to scale)

The four outer planets, or gas giants, (sometimes called Jovian planets) are so large that they collectively make up 99 percent of the mass known to orbit the Sun. Jupiter and Saturn are true giants, at 318 and 95 Earth masses, respectively, and composed largely of hydrogen and helium. Uranus and Neptune are substantially smaller, being only 14 and 17 Earth masses, respectively. Their atmospheres contain a smaller percentage of hydrogen and helium, and a higher percentage of "ices," such as frozen water, ammonia, and methane. For this reason some astronomers have suggested putting them in a separate category"Uranian planets" or "ice giants." All four outer planets exhibit orbital debris rings, although only the ring system of Saturn is easily observable from Earth. The term *outer planet* should not be confused with *superior planet*, which designates all planets that lie outside Earth's orbit (thus consisting of the outer planets plus Mars).

Jupiter

Jupiter (5.2 AU), at 318 Earth masses, is 2.5 times the mass of all the other planets put together. Its composition of largely hydrogen and helium is not very different from that of the Sun. Jupiter's strong internal heat creates a number of semi-permanent features in its atmosphere, such as cloud bands and the Great Red Spot. Three of its 63 satellites Ganymede, Io, and Europa share elements in common with the terrestrial planets, such as volcanism and internal heating. Ganymede has a larger diameter than Mercury.

Saturn

Saturn (9.5 AU), famous for its extensive ring system, has many qualities in common with Jupiter, including its atmospheric composition, though it is far less massive, being only 95 Earth masses. Two of its 49 moons, Titan and Enceladus, show signs of geological activity, though they are largely made of ice. Titan, like Ganymede, is larger than Mercury; it is also the only satellite in the solar system with a substantial atmosphere.

Uranus

Uranus (19.6 AU), at 14 Earth masses, is the lightest of the outer planets. Uniquely among the planets, it orbits the Sun on its side; its axial tilt lies at over 90 degrees to the ecliptic. Its core is remarkably cooler than that of the other gas giants (though it is still several thousand degrees Celsius) and radiates very little heat into space. Uranus has 27 satellites, the largest being Titania, Oberon, Umbriel, Ariel, and Miranda.

Neptune

Neptune (30 AU), though slightly smaller than Uranus, is denser and slightly more massive, at 17 Earth masses. It radiates more internal heat than Uranus, but not as much as Jupiter or Saturn. Its peculiar ring system is composed of a number of dense "arcs" of material separated by gaps. Neptune has 13 moons. The largest, Triton, is geologically active, with geysers of liquid nitrogen.

Edgeworth-kuiper Belt

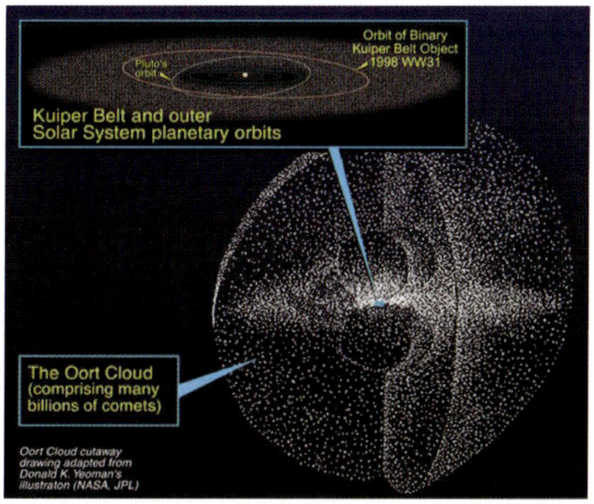

Artist's rendering of the Kuiper Belt and hypothetical Oort cloud

The area beyond Neptune, sometimes referred to as the "outer solar system" or the "trans-Neptunian region," is still largely unexplored.

This region's first formation, which actually begins inside Neptune's orbit, is called the Edgeworth-Kuiper belt (or simply, Kuiper belt). This belt is a great ring of debris similar to the asteroid belt, but it is composed mainly of ice and is far greater in extent, lying between 30 and 50 AU from the Sun. This region is thought to be the place of origin for short-period comets, such as Halley's comet. Though it is composed mainly of small solar system bodies, many of the largest Kuiper belt objects could soon be reclassified as dwarf planets. There are estimated to be over 100,000 Kuiper belt objects with a diameter greater than 50 kilometers; however, the total mass of the Kuiper belt is relatively low, perhaps barely equaling the Earth's mass. Many Kuiper belt objects have multiple satellites, and most have orbits that take them outside the plane of the ecliptic.

The Kuiper belt can be roughly divided into two regions:

1. The "resonant" belt, consisting of objects whose orbits are in some way linked to that of Neptune (orbiting, for instance, three times for every two Neptune orbits, or twice for every one). It actually begins within the orbit of Neptune itself.
2. The "classical" belt, consisting of objects that don't have any resonance with Neptune. It extends from roughly 39.4 AU to 47.7 AU.

Pluto and Charon

Pluto, Charon, and their two known moons

From 1930 to 2006, Pluto (39 AU average), one of the largest known Kuiper belt objects, was considered the solar system's ninth planet. In 2006, however, the International Astronomical Union (IAU) created an official definition of a planet that disqualified Pluto as a planet. As a result, Pluto was reclassified as a dwarf planet. Pluto has a relatively eccentric orbit, inclined 17 degrees to the ecliptic plane and ranging from 29.7 AU from the Sun at perihelion (within the orbit of Neptune) to 49.5 AU at aphelion. Prior to the 2006 redefinitions, Charon was considered a moon of Pluto, but in light of the redefinition, it is unclear whether Charon will continue to be classified as Pluto's moon or as a dwarf planet itself. Charon does not orbit Pluto rather, both bodies orbit a barycentre of gravity in empty space, making Pluto-Charon a binary system. Two much smaller moons, Nix and Hydra, orbit Pluto and Charon.

Kuiper belt objects that, like Pluto, possess a 3:2 orbital resonance with Neptune that is, they orbit twice for every three Neptunian orbits are called Plutinos. Other Kuiper belt objects have different resonant orbits (such as 2:1, 4:7, 3:5) and are grouped accordingly. The remaining Kuiper belt objects, in more "classical" orbits, are classified as Cubewanos.

Comets

Comets are small solar system bodies (usually only a few kilometers across) containing large amounts of volatile ice and tracing highly eccentric orbits. They generally have a perihelion within the orbit of the inner planets and an aphelion far beyond Pluto. When a comet approaches the Sun, its icy surface begins to sublimate, or boil away, creating a coma a long tail of gas and dust that is often visible with the naked eye.

Comet Hale-Bopp

There are two basic types of comets:

1. Short-period comets, with orbits less than two hundred years
2. Long-period comets, with orbits lasting thousands of years

Short-period comets are believed to originate in the Kuiper belt, while long period comets, such as Hale-Bopp (pictured), are thought to originate in the Oort cloud. Some comets with hyperbolic orbits may originate outside the solar system. Old comets that have had most of their volatiles driven out by solar warming are often categorized as asteroids (likely to be D-type asteroids).

Centaurs are icy, comet-like bodies that have less eccentric orbits, so that they remain in the region between Jupiter and Neptune. The first centaur to be discovered, 2060 Chiron, has been called a comet because it was found to develop a coma just as comets do when they approach the Sun.

Scattered Disc

Black: scattered disc; blue: classical Edgeworth-Kuiper belt; green: resonant Edgeworth-Kuiper belt objects, including Pluto

Overlapping the Kuiper belt but extending much further outward is what is called the scattered disc. Scattered disc objects are believed to have been originally native to the Kuiper belt but were ejected into erratic orbits in the outer fringes by the gravitational influence of Neptune's outward migration. Most scattered disc objects have perihelia within the Kuiper belt but aphelia as far as 150 AU from the Sun. Their orbits are also highly inclined to the ecliptic plane, and are often almost perpendicular to it. Some astronomers, such as Kuiper belt co-discoverer David Jewitt, consider the scattered disc to be merely another region of the Kuiper belt, and they describe scattered disc objects as "scattered Kuiper belt objects."

Eris

Eris (68 AU average) is the largest known scattered disc object. It was the cause of the most recent debate about what constitutes a planet, because it is at least five percent larger than Pluto, with an estimated diameter of 2,400 kilometers (1,500 miles). It is now the largest of the known dwarf planets. It has one moon named Dysnomia.

This object has many similarities with Pluto. Its orbit is highly eccentric, with a perihelion of 38.2 AU (roughly Pluto's distance from the Sun) and an aphelion of 97.6 AU. Also, the orbit is steeply inclined to the ecliptic plane, at 44 degrees, more so than any known object in the solar system except for the newly discovered object 2004 XR_{190}(also known as "Buffy"). It is believed to consist largely of rock and ice.

Farthest Regions

The point at which the solar system ends and interstellar space begins is not precisely defined, because the solar system's outer boundaries are delineated by two separate forces: the solar wind and the Sun's gravity. The solar wind extends to a point roughly 130 AU from the Sun, whereupon it surrenders to the surrounding environment of the interstellar medium. It is generally accepted, however, that the Sun's gravity holds sway over the hypothetical Oort cloud. This cloud, conceived of as a great mass of up to a trillion icy objects, is believed to be the source of all long-period comets. It is thought to surround the solar system like a shell from 50,000 to 100,000 AU beyond the Sun, or almost a quarter of the distance to the next star system. From this perspective, the vast majority of the solar system is completely unknown. Nonetheless, recent observations of the Solar system and other star systems have led to an increased understanding of what is or may be lying at its outer edge.

Sedna

An artist's conception of Sedna

Sedna is a large, reddish Pluto-like object with a gigantic, highly elliptical orbit that takes it from about 76 AU at perihelion to 928 AU at aphelion and takes 12,050 years to complete. Mike Brown, who discovered the object in 2003, asserts that it cannot be part of the scattered disc or Kuiper Belt, as it has too distant a perihelion to have been affected by Neptune's migration. He and other astronomers consider it to be the first in an entirely new population one that may include the object 2000 CR_{105}, which has a perihelion of 45 AU, an aphelion of 415 AU, and an orbital period of 3,420 years. Sedna is very likely a dwarf planet, though its shape has yet to be determined with certainty.

Heliopause

The solar wind, a stream of ions, electrons, neutral matter, and magnetic fields emanating in great pulses from the sun, sustains the heliosphere, which is thought to be slightly teardrop shaped and to extend out from the sun to about 95 AU, or two and one half times the orbit of Pluto. The bubble's diaphonous edge begins at the termination shock the point at which the solar wind is significantly slowed by collision with opposing plasma winds of the interstellar medium. Here the wind condenses and becomes more turbulent in a region known as the heliosheath, which extends outward for a further 40 AU at its stellar-windward side, but stretches many times that distance in the opposite direction. The outer boundary of the sheath, the heliopause, is the region in which the solar wind is fully overcome by the interstellar galactic wind. Beyond the heliopause, at around 230 AU, the interstellar plasma registers the approach of the solar system in the form of the bow shock a plasma "wake" generated by the Sun as it travels through the Milky Way.

Galactic Context

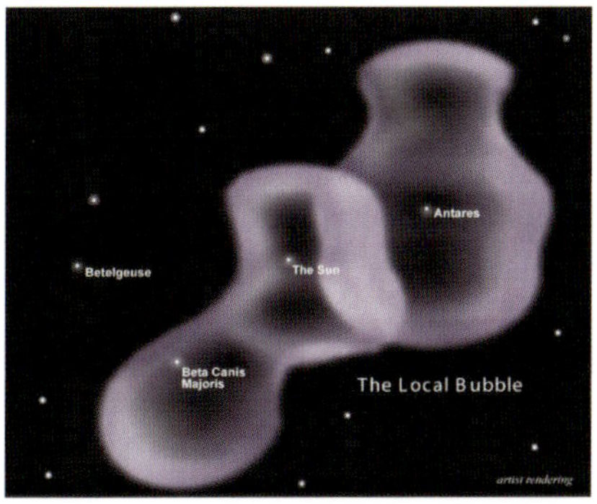

Artist's conception of the Local Bubble

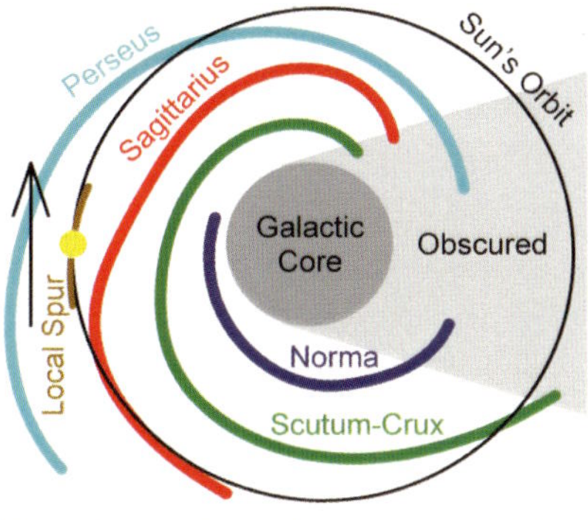

Presumed location of the Solar system within our galaxy

The solar system is located in the Milky Way galaxy, a barred spiral galaxy with a diameter estimated at about 100,000 light years containing approximately 200 billion stars. Our Sun resides in one of the Milky Way's outer spiral arms, known as the Orion Arm or Local Spur. The immediate galactic neighbourhood of the solar system is known as the Local Fluff, an area of dense cloud in an otherwise sparse region known as the Local Bubble, an hourglass-shaped cavity in the interstellar medium, roughly three hundred light years across. The bubble is suffused with high-temperature plasma that suggests it is the product of several recent supernovae.

Estimates place the solar system at between 25,000 and 28,000 light-years from the galactic centre. Its speed is about 220 kilometers per second, and it completes one revolution every 226 million years. It appears to have a very remarkable orbit one that is both extremely close to being circular, and at nearly the exact distance at which the orbital speed matches the speed of the compression waves that form the spiral arms.

The apex of solar motion that is, the direction in which the Sun is heading is near the current location of the bright star, Vega. At the galactic location of the solar system, the escape velocity with regard to the gravity of the Milky Way is about 1,000 kilometers per second.

Extrasolar Planets

Until 1994, the solar system was the only known example of a stellar system in which a planet or planets orbit around a star. Since that first discovery, the subsequent discovery of more than 200 extrasolar planets has led to the term "solar system" being applied generically to all stellar systems. Technically, however, it should strictly refer to Earth's system only, as the word "solar" is derived from the Sun's Latin name, *Sol*. Other stellar systems or planetary systems are usually referred to by the names of their parent star, such as "the Alpha Centauri system" or "the 51 Pegasi system."

Studies of extrasolar systems have shown that the vast majority discovered so far are markedly different from the solar system. For instance, many extrasolar planetary systems contain a "hot Jupiter" a planet of size comparable to Jupiter, that nonetheless orbits very close to its star (for instance, at 0.05 AU). It has been hypothesized that while the giant planets in these systems probably formed in the same relative place as the gas giants in Earth's solar system did, the extrasolar giant planets may have migrated inward toward their parent star and along the way caused the destruction or ejection from the system of any existing smaller planets. The apparent prevalence of hot Jupiter's among the extrasolar planets discovered so far may be an artifact of a skewed data set arising simply because close-in giant planets are the easiest extrasolar planets to detect. In coming years, it is quite possible that ongoing improvements in the capabilities for detecting extrasolar planets will lead to the discovery of stellar systems much more similar to our own.

Discovery and Exploration

For many thousands of years, most people thought of the Earth as a stationary body at the centre of the universe, and categorically different from the ethereal objects that moved through the sky. The conceptual advances of the seventeenth century led by Nicolaus Copernicus, Galileo Galilei, Johannes Kepler, and Isaac Newton gradually guided people to accept ideas such as (a) the Earth moves in an orbit around the Sun, (b) the planets are governed by the same laws as the Earth, and (c) the planets could therefore be similar to the Earth.

Telescopic Observations

When astronomers first began using the telescope, they made unexpected discoveries about the solar system. For instance, Galileo Galilei discovered that the Moon was cratered, that the Sun was pocked with sunspots, and that Jupiter had four satellites in orbit around it. Giovanni Domenico Cassini and Christian Huygens followed on from Galileo's discoveries by discovering the rings of Saturn, and Saturn's moon Titan, respectively.

In 1682 Edmund Halley realized that repeated sightings of a comet were in fact recording the same object, returning regularly once every 75-76 years. This proved that comets were not atmospheric phenomena, as was previously thought, and provided the first evidence that objects other than planets orbited the Sun.

In 1781 William Herschel was looking for binary stars in the constellation of Taurus when he observed what he thought was a new comet. In fact, its orbit revealed that it was a new planet, Uranus, the first ever discovered with a telescope. In 1801 Giuseppe Piazzi discovered Ceres, a small world between Mars and Jupiter, that was initially considered a new planet. Subsequent discoveries of thousands of other small worlds in the same region led to their eventual reclassification as asteroids.

In 1846 discrepancies in the orbit of Uranus led many to suspect a large planet must be tugging at it from farther out. Urbain Le Verrier's calculations eventually led to the discovery of Neptune.

Further discrepancies in the orbits of the planets led Percival Lowell to conclude that yet another planet, "Planet X," must still be out there. After his death, his Lowell Observatory conducted a search, which ultimately led to Clyde Tombaugh's discovery of Pluto in 1930. Pluto was, however, found to be too small to have disrupted the orbits of the outer planets, and its discovery was therefore coincidental. Like Ceres, it was initially considered a planet, but after the discovery of many other similarly sized objects in its vicinity, it was eventually reclassified as an Edgeworth-Kuiper belt object.

Observations by Spacecraft

Planets and Small Rocky Bodies

Since the start of the space age, many explorations have been performed by unmanned space missions that have been organized and executed by various space agencies. The first probe to land on another solar system body was the Soviet Union's Luna 2, which impacted the Moon in 1959. Since then, increasingly distant planets have been reached, with probes landing on Venus in 1965, Mars in 1976, the asteroid 433 Eros in 2001, and Saturn's moon Titan in 2005. Spacecraft have also made close approaches to other planets for instance, Mariner 10 passed Mercury in 1973.

The first probe to explore the outer planets was Pioneer 10, which flew by Jupiter in 1973. Pioneer 11 was the first to visit Saturn, in 1979. Following their launch in 1977, Voyagers 1 and 2 performed grand tours of the outer planets, with both probes passing Jupiter in 1979 and Saturn in 1980-1981. Voyager 2 then went on to make close approaches to Uranus in 1986 and Neptune in 1989. The Voyager probes are now far beyond Neptune's orbit, and astronomers anticipate that they will encounter the heliopause, which defines the outer edge of the solar system, in the next few years.

All planets in the solar system have been visited to varying degrees by spacecraft launched from Earth, the last being Neptune in 1989. Through these unmanned missions, humans have been able to get close-up photographs of all the planets and, in the case of landers, perform tests of the soils and atmospheres of some.

No Kuiper belt object has been visited by a manmade spacecraft. Launched on January 19, 2006, the *New Horizons* became the first manmade spacecraft designated to explore this area. This unmanned mission is scheduled to fly by Pluto in July 2015. Should it prove feasible, the mission will then be extended to observe a number of other Kuiper belt objects.

Plasma

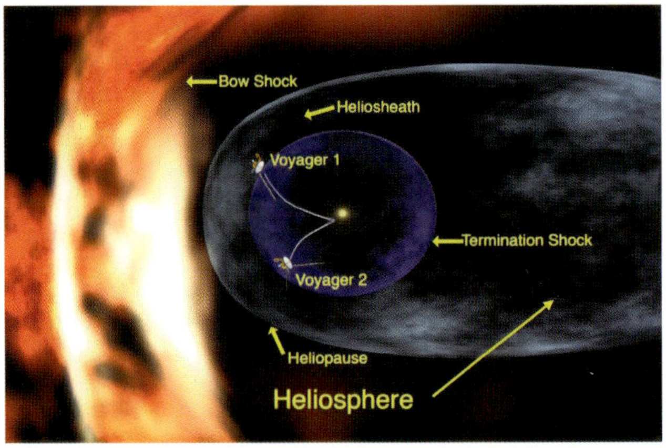

Voyagers 1 and 2 entering the heliosheath

From the beginnings of space exploration, spacecraft have revealed magnetic fields, charged particles and electrical activity along with neutral atoms in what had been assumed to be empty space. Discovery of the Van Allen radiation belts surrounding the earth were an early clue that space was much more dynamic than could be known from the surface of the planet. Beyond these belts, spacecraft discovered a pulsing plasma stream, which was named the solar wind, and intermediate between the belts and the wind, earth's magnetopause where the influence of earth's magnetic field succumbed to the sun's magnetic field carried in the solar wind.

The story of the exploration of the solar system is incomplete without an appreciation of the extent to which exploration spacecraft are engaged in monitoring magnetic fields, charged particles, and current flows both in the vicinity of planets and in the great voids between them. In all their journeys, the exploration spacecraft have always detected the solar wind still streaming away from the sun. An important realization for life on planet earth is that the sun supports it by providing not only heat and light, but also protection from many potentially harmful cosmic rays (extremely high energy particles), which are diverted from deep penetration of the solar system by the plasma envelope of the heliosphere sustained by the solar wind.

At Home in the Solar System

To ancient peoples the sky was the home of gods and goddesses associated with what we today know as the Sun, the Moon, and the visible planets. Tracking the activities, the movements in the sky, of these deities was important for planning and conducting life activities, as evidenced by astrological systems practiced for centuries or even millenniums in places such as China, India, Mesopotamia, Egypt, Greece, Rome, and Central America. Each of these accorded central importance to the visible planets plus the Sun and the Moon, and several of these systems are still applied today. They were developed based on the assumption that the earth was the centre about which the gods and goddesses moved through the sky.

In modern times, monotheistic religions and science have together severely undermined the mystery and primacy of the deities in the sky. Conceptually, the monotheistic religions have denied their existence, while science has explained them away and replaced them with inconceivably remote and totally lifeless bodies of plasma (the Sun), rock (the Moon, Mercury, Venus, and Mars), or gas (Jupiter and Saturn). The birth of the solar system as a distinct conceptual unit stripped of its mysterious gods and influence on human affairs traces back to the radical shift some four centuries ago from the earth-centred model of the cosmos to the sun-centred solar system cosmos.

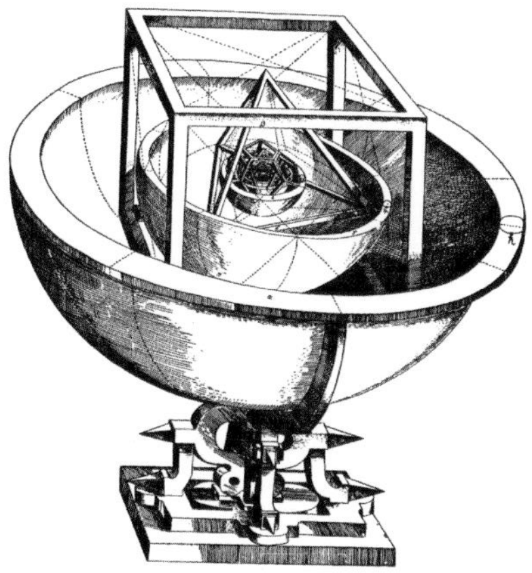

Kepler's Platonic solid model of the solar system from *Mysterium Cosmographicum* (1596)

One of the towering figures in that transition, the German astronomer Johannes Kepler, a Lutheran and still committed to the Aristotelian idea that the planets were moving on clear spheres, devised a model in which each planet moved on a sphere either circumscribed around or inscribed within one of the five Platonic solids, which were nested in an order that yielded orbital dimensions strikingly close to those determined by the astronomical methods available at the time. Kepler's proposal was one of the last serious attempts at producing a model of planetary motions that retained explicit association in this case, through the Platonic ideal forms with a higher realm of mystery and deity.

The solar system given to us by science, despite its lack of deity and mystery, has nonetheless claimed a new place of meaning in the human mind as an archetypal model of organization that can be applied beyond the realms of astronomy. This model is one of relationships in which two or more elements sharing a common purpose interact as partners, with the dominant element assuming the position of the central subject partner while the other element or elements follow the lead of (orbit around) the subject and they all together create something new. For example, the solar system, which his given birth to both life and human beings, is something new that the planets together with the sun and all the moons are co-creating.

In the early twentieth century, the first effective working model of the hydrogen atom resembled a simplified miniature solar system with one "planet," an electron, orbiting at a great distance around a massive, central "sun," a proton. Although the resemblance was forced and the model was rather quickly replaced by a quantum shell model, the concept of a tiny solar system in an atom has embedded itself in society in the icon representing an atom and atomic energy; and the solar system model is still used for elementary instruction about atoms.

Many scientists decry the continued association of the atom with the solar system model because quantum models of the atom present a nucleus that is not orbited by electrons but rather is surrounded by a nebulous cloud of quantized probable electron locations described not as an orbit but as an orbital. Disregarding or remaining ignorant of such concerns, those who view the solar system as an archetype persist in seeing the atom as embodying the same archetype when they consider that the tiny, mobile electrons and the massive, dominant nucleus collaborate in producing a new creation, the atom.

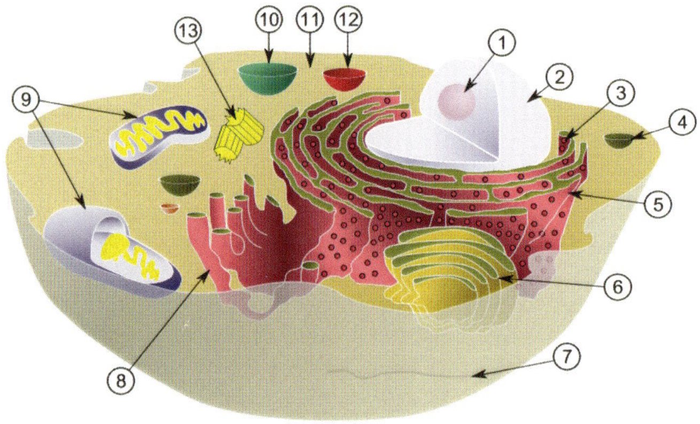

With a dominant nucleus (2) surrounded by many collaborating organelles, a typical animal cell shown here can be seen as embodying the same archetype as the solar system with its dominant Sun and orbiting planets

A similar archetypal view can be applied as well to biological cells with their dominant nucleus orchestrating the amazingly complex metabolism carried out in the surrounding cytoplasm in a harmony of action that produces and maintains a functioning cell.

Similarly, within the human nuclear family, the parents and the children can be seen as occupying positions comparable respectively to those of the sun and the planets. Further, the parents in giving love, nurture, and guidance to the children are acting on the model of the sun gravitationally directing the earth while also providing earth with the heat and light essential for generating and sustaining life.

Viewed from yet another angle, the vertical order of the extended family generations running back from parents to grandparents and great grandparents and forward to children and grandchildren is similar to the astronomical vertical order mapped by taking the Sun to be in the position of the parents. "Generations" extending "back" from the Sun would be the centre of the Milky Way galaxy about which the Sun orbits, and beyond that the massive supercluster pulling the Milky Way toward it. Generations extending "forward" from the Sun would be the planets in the position of children and the moons of planets in the position of grandchildren. Seeing such a harmony of pattern extending from the solar system to the family, the cell, and the atom can affirm the human sense of the connectedness of all things and of the solar system as home for the human family.

Formation and Evolution of the Solar System

The formation and evolution of our solar system (and planetary systems around other stars) are among the most challenging and intriguing fields of modern science. As the product of a long history of cosmic matter evolution, this important branch of astrophysics is referred to as stellar-planetary cosmogony. Interdisciplinary by way of its content, it is based on fundamental theoretical concepts and available observational data on the processes of star formation. Modern observational data on stellar evolution, disc formation, and the discovery of extrasolar planets, as well as mechanical and cosmochemical properties of the solar system, place important constraints on the different scenarios developed, each supporting the basic cosmogony concept (as rooted in the Kant-Laplace hypothesis). Basically, the sequence of events includes fragmentation of an original interstellar molecular cloud, emergence of a primordial nebula, and accretion of a protoplanetary gas-dust disk around a parent star, followed by disk instability and break-up into primary solid bodies (planetesimals) and their collisional interactions, eventually forming a planet.

Planets formation is a widespread although very complex process, believed to be the succession of several stages affected by different mechanisms of physical interactions, chemical transformations, and numerous perturbations in the gas-dust disk. Scenarios and model approach to the origin of protoplanetary nebulae and evolution are generally backed by observational data. The mechanical, astrophysical, and cosmochemical characteristics of the solar system serve as the starting concept for the formation of planets around stars. The solar system planets and satellites architecture, as well as existing patterns in the systems of extrasolar planets definitely point to a unified process of every system formation though with different constraints. Data available on surface properties and matter composition for the solar system planets when confronting the samples of material from their embryos (small bodies) and "debris" (meteorites) provide an insight into the probable sources, paths, and chronology of this process.

Cosmochemistry and Chronology of Evolution

Of primary importance is an opportunity of getting insight into the chronology of the key physical and chemical mechanisms responsible for the early solar system evolution. Study of meteorites is the main tool of cosmochemistry aimed to reconstruct the processes of matter origin and transformations in the protoplanetary disc and forming bodies.

The time sequence was established based on the measurements of ratios of long- and short-lived isotopes and products of their decay in meteoritic materials. The main isotopic systems used in the study were U, Th-Pb, Sm-Nd, Al-Mg, Mn-Cr, Rb-Sr, I-Xe, Hf-W. Many undifferentiated meteorites (chondrites) contain the refractory inclusions of microns to cm in size enriched in refractory elements such as Al and Ca (Calcium Aluminum Inclusions or CAIs). They were assumed to belong to the ancient solid material that condensed out near the Sun ($r < 0.5$ AU) at $T \sim 2000–1700$ K. These objects, including some ultra-refractory mineral nodules, enabled a determination of the absolute age of the solar system. The measured values vary from 4567.1 ± 0.1 Ma to 4568.67 ± 0.17 Ma, with the most reliable being 4567.30 ± 0.16 Ma. Thus, the time of the solar system origin is determined with accuracy of better than ~1 Ma, or 0.02%. Concurrently, the absolute age of iron and stony meteorites of different petrological classes was defined from 1 to a few Ma younger CAI. Let us note that the submillimeter chondrules (*spherules*) embedded in stony meteorites and composed of ferromagnesian silicates are dated in the range from 4567.32 ± 0.42 to 4564.71 ± 0.30 Ma indicating an age gap between CAIs and chondrules with implication that chondrules formation lasted ~3 Ma. This time scale is in accord with protoplanetary disk lifetimes inferred from astronomical observations.

One may assume that during a few million years, interval accumulation and thermal evolution (differentiation) of the parent bodies of these ancient meteorites occurred. Provisionally the first primordial parent bodies of ~100 km in size formed in the very first few million years since the solar system origin. Such a size was sufficient for the body to experience differentiation due to intense heating by the short-lived ^{26}Al and ^{60}Fe nuclides with an iron core emergence. The subsequent core and silicate shell fragmentation caused by numerous collisions have been probably responsible for the existing iron and stony meteorites. Otherwise their existence is difficult to explain, in contrast to non-differentiated chondrites that experienced no melting by the exhausted short-lived isotopes heat source.

The above time scale is in accord with the results of computer modeling that argue that accretion of matter from the disk on the protosun terminated in 1–2.5 Ma after the system formation. The dust subdisk composed presumably of 1–10 cm particles is believed to form much earlier, in 0.01–0.1 Ma at radial distance r ~1 AU. Here critical density was achieved and gravitational instability developed. Evidently, this time was sufficient for accumulation and thermal evolution of the first solid bodies. Assuming that mass of the protoplanetary disc M_d was ~0.1 M_S and that with account for the disc partial dissipation ~0.1 M_d ultimately entered the planets, we may estimate ~10^{10} of ~100 km original bodies were born in the first ~2 Ma. This idea is in accord with the models favoring distribution of asteroids from the initial generation of planetesimals of nearly similar size on which chondrules have been presumably accreted.

Further Advancement and Current State

We shall now discuss the state of the art in our views on principal mechanisms of the solar system origin. In modern astronomy, the key consists of high resolution images and spectral features of objects relevant to planets formation at the different stages of evolution. In computer modeling, the focus is given to the theoretical treatment and development of robust models and effective algorithms allowing us to get insight into genesis of planetary system origin from primordial matter of the outer space involving disk formation, its radial/vertical compression, and dust distribution/size grow affecting the disc structure. Unfortunately, unlike protoplanetary accretion disks whose structure and evolution are accessible to astronomical observations, the mechanism of primary solid bodies set up in the gas-dust disk and their growing to planetary embryos remains rather speculative because cannot be yet tested experimentally. Hence, computer modeling is essentially the only tool to reconstruct the multiple processes involved with the use of observational constraints for models verification.

Astronomical Observations

Generally, our current understanding of the mechanism of solar system origin is in accord with the experimental data available. The discovery of circumstellar disks through high-resolution visual, infrared, and submillimeter observations, and especially extrasolar planets, extend considerably our views of the discs structure and properties of emerging planetary systems with a variety of configurations. Space telescopes such as Hubble, Spitzer, and Herschel provided an exciting picture of how all the ingredients of the cosmic stew that makes planetary systems from a protostar nebula are blended together. The concept needs, however, more rigorous and augmented support and this is the focus of the progressively refined models with an ultimate goal to reconstruct a scenario of evolution and the processes involved.

The fact that a significant fraction of young stars is surrounded by disks had already become obvious by the early 1990s, though historically the first disks were discovered around stars more massive than the Sun, such as Vega (α Lyra). The observations of young protostellar and stellar objects are currently performed over a very wide wavelength range: from X-rays to the radio band. One of the most informative methods for studying these objects is to analyze their spectral energy distributions. The studies of the infrared, submillimeter, and millimeter spectra have revealed gas-dust disks with Keplerian rotation around hundreds of T Tauri stars (young stars of less than 2 solar masses) with ages ranging from 10^5 to 10^7 yr. Gas-dust disks have been discovered around most of the observed T Tauri stars with ages 10^6 yr and around ~20–30% of the stars with ages 10^7 yr with a mean disk lifetime of 3–6 Ma. The disk masses turned ~0.01–0.2 M_O while their extent up to ~100 AU.

Nowadays it is recognized that both young (10^6–10^7yrs) and older (10^7–10^8yrs) stars possess disks either before entering or already residing on the Main Sequence of the H-R diagram. However, young stars (of T-Tauri type) are grouped in the regions of star formation (such as Orion Nebula) where they dominate (> 70–80%) and contain more gas and dust than the older ones. Obviously, this is because the process of planet formation completed in the older disks and respectively, the major part of matter is in the solid bodies while gas was lost. Dust in these disks appears to have secondary origin as the result of numerous collisions and debris left behind after planets formation and is partially depleted by accretion on growing bodies.

Thermal infrared gas and dust emission measurements and high-angular resolution observations in submillimeter and millimeter wavelength made the most significant contribution to the study of disks' structure and chemistry, though the data interpretation is not straightforward. They revealed that the bulk of the matter is concentrated in an inner region with a radius of ~40 AU (inner boundary of Kuiper Belt Objects KBO locations in the solar system) and this similarity of parameters per se suggests that these disks with ages 10^6–10^7 yrs are probable precursors of planetary systems (Beckwith & Sargent, 1996; Cieza, Kessler-Silacci, Jaffe, Harvey, & Evans, 2005). The accretion rate (total mass flux) from the disk onto the central star to be estimated: for most stars, it lies within the range $\dot{M}$ ~10^{-9}–$10^{-7}M_O$ /yr with a mean value of ~10^{-8} M_O/yr; in the range of stellar ages from 10^5 to 10^7, there is a tendency for the flux to decrease from 10^{-7} to $10^{-9}M_O$ /yr (Calvet, D'Alessio, Hartmann, Wilner, Walsh, & Sitko, 2002; Hueso & Guillot, 2005). A rather low gas content in the range of 10^{-2} to 10^{-3} disk mass was found indicating low gas to dust mass ratio of the order of 10 in contrast to an earlier estimate of nearly an order of magnitude higher.

Many molecular species were identified and their isotopic compositions were measured. Numerous molecules found in protoplanetary disks (such as H_2O, CO, N_2, H_2CO, HCN, etc.) are probably genetically related to the volatiles contained in the frozen granules of primordial accreting material. They are supposed to be subsequently subjected to substantial chemical and thermal processing. Some of these molecules are apparently ionized or in non-equilibrium state due to the photolysis attributable to far ultraviolet and X-ray radiation from a young star. Observations carried out with the Spitzer Space Telescope allowed the discovery of several young stars surrounded by disks in a fairly small star-forming region. The size of some disks is comparable with the Neptune orbit in the solar system figure.

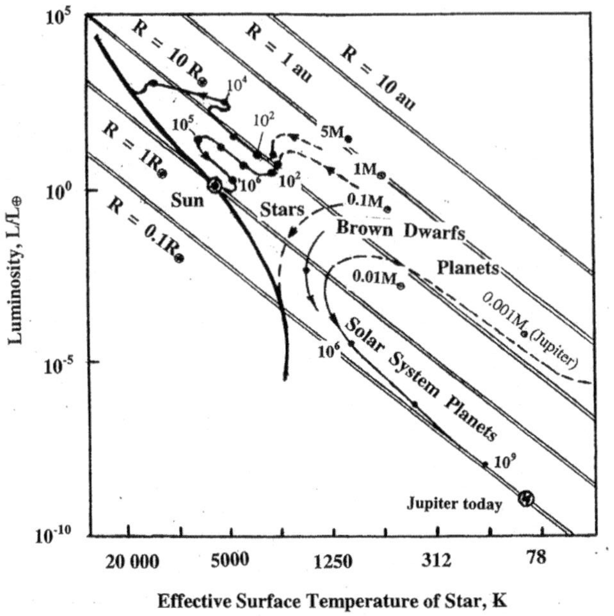

Figure: A part of the Hertzsprung-Russell diagram. Diagonal lines running from the upper left to the bottom right are disc radii from 10 AU to 0.1_O. The solid curve represents the main sequence. Tracks for protostars of the solar (larger and lower) masses are shown. Objects having $M << M_O$ do not reside in the MS

Note that the Kuiper belt could be regarded as analogous to the low-mass dusty debris disks observed around many Main Sequence stars where existence of planets is evidenced by some disk warping.

Of great importance for understanding evolution of the protoplanetary accretion disk, in particular its inhomogeneous structure, thermal regime, and dynamics of its inner regions, are the data on dust composition and evolution. Observations were carried out in the optical, near, and thermal infrared spectral ranges and included measurements of emission spectra with the long baseline interferometry in the millimeter wavelength. Many specific features in the thermal gas-dust emissions were revealed, including sources for chemical processing of disc gaseous matter and finding spectacular dust rings. The physical characteristics and mineralogy of dust particles were inferred assuming a close analogy with meteor particles in the Earth's atmosphere. Interestingly, disk particles turned out to be much larger than micron-size dust in the diffuse interstellar medium: The largest are millimeters to centimeters across and resemble sand or even pebbles. Moreover, they exhibit the height stratification, with the smaller micron-size particles being concentrated near the disk surface. Such stratification is thought to persist for millions of years. Naturally, the content and size distribution of solid particles (granules) affects the disc medium opacity and turbulence flow patterns. They strongly influence the disk thermal regime, viscous properties, chemical transformations in a gaseous medium and, in the end, its evolution including the processes' dependence on the radial distance from the protosun and the early subdisk formation.

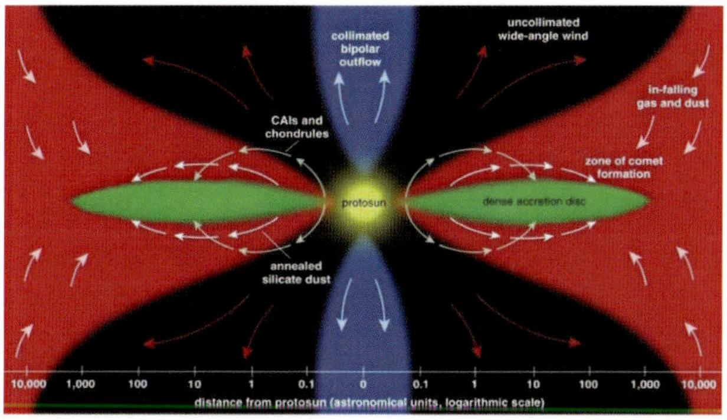

Figure: Scheme for the formation of a gas-dust accretion disk and a subdisk.

The proto-Sun onto which matter from the protoplanetary nebula (the red color) continues to accrete is at the centre. The green color indicates the forming dust subdisk, near which the outflow of gas and dust, including the formed high-temperature condensates in the inner zone, such as refractory CAIs, takes place. The blue color indicates the bipolar flows of matter attributable to the solar magnetic field.

Based on observed particles' properties, it may be reasonably assumed that their genesis is related to the formation very close to protostar and that some of them may have undergone the follow-up processes of evaporation-crystallization during radial motion in the disk. Besides, they may experience heating by shock waves in the accretion zone and subsequent rapid cooling. In accord with this concept is the finding the same type of rather refractory materials in comets, in particular those coming from the frigid outskirts of our solar system. Comets are known to have been born in regions beyond the line where water becomes frozen (called the snow line) and are referred to as remnants of the icy planetesimals. A probable scenario one may assume is that comets formed due to gravitational collapse of pebble clouds emerged through the streaming instability, which acquired the solids that would have been frozen into comets in the planet-forming disk regions.

Computer and Lab Modeling

The matter of the protoplanetary gas-dust disk is a complex system of the different phase composition, densities, temperatures, and degrees of ionization, which vary with radial distance. Basically, it is an inhomogeneous medium composed of gas and dust particles of various sizes and origin. This matter, which is generally magnetized dusty plasma, is in a state of turbulence depending on the radial and azimuthal position of a parcel of matter.When the main dynamical forces controlling the rotating disc flattening (gravitational and centrifugal) are in balance, weaker factors, such as the thermal/viscous processes, turbulence, and electromagnetic phenomena dominate the disk's evolution. They certainly affect the condensation of volatiles, including first of all water, and bear significant effect on the relative content and abundance of gaseous species and solid particles, as well as disk energetic and angular momentum transport.

When the plasma effects are disregarded, the motion of a disk medium containing dust suspended in gas can be modeled most adequately within the framework of mechanics of heterogeneous turbulized media with allowance made for the physical-chemical properties of the phases, heat and mass transport, incident radiation/opacity changes, viscosity variations, chemical reactions, phase transitions, coagulation, fragmentation, etc. Specifically, it is focused on the dynamical interaction of turbulized gas and dust including modification of the turbulence energy of the carrier phase by solid particles (i.e., the reverse effect of the dust component on the turbulent and thermal regimes of the disk gas component); the influence of turbulence on the rates of phase transitions (evaporation, condensation); on the jump-like disperse particle accumulation processes such as coagulation and fragmentation during mutual collisions between particles in the mass flow; and, finally, on the settling of solid particles through the gas to the disk midplane, where they form a flattened dust layer a geometrically thin subdisk.

Obviously, the presence of polydisperse (different particles' size) admixture in a turbulized medium complicates significantly the disk hydrodynamics, contributing to the realization of additional regimes of cosmic matter flow. Note that the synergetic collective self-organization processes in the thermodynamically open system of the protoplanetary disk against the background of a large-scale shear flow of cosmic matter associated with its differential rotation are regarded as very important mechanisms shaping the properties of a viscous accretion disk at various stages of its evolution.

Whatever is the character of events under consideration, it is clear that complex physical and chemical processes accompanying evolution of the heterogeneous medium where dust particles collisions domain, are responsible for the first solids origin and planetesimal's formation. The developed models include the sequence of changes in the aggregate state of the main protoplanetary matter components; the location of the condensation-evaporation fronts depending on the thermodynamic parameters of the disk; the role of particle sublimation and coagulation in the two-phase medium with the account for particle size distribution; the relative contribution of radiation and turbulence to the heat and mass transport; and the mechanisms for the development of streaming and gravitational instabilities with allowance made for the shear stresses in boundary layers and polydispersed, suspended dust particles.

In the most comprehensive approach, a continuum model of heterogeneous disk medium should take into account the joint influence of MHD effects and turbulence on the dynamics and heat and mass transport processes in differentially rotating matter with allowance made for the inertial properties of the polydispersed admixture of solid particles, coagulation, radiation, and changing partitioning of elements between gaseous and condensed phases. Turbulence generated at the boundaries of the protoplanetary disk layers and caused by shear flows corresponds in character to the parameters of a boundary (Ekman) layer and significantly affects the disc dynamics including the Kelvin-Helmholtz instability. It is important to emphasize that generation and maintenance of shear turbulence at various evolutionary stages of the disk involves a two-phase (gas-dust) medium with a differential angular velocity of rotation, different relative contents of dust particles, their size distribution, and coagulation processes. In general, a heterogenic mechanics approach should be applied to account for the emergence of coherent order against the background of random motions in large-scale turbulent structures. Also, the evolution of turbulence in the rotating accretion disk is supposed to be influenced by hydrodynamic helicity responsible for the cascade process of the inverse energy transfer from small to large eddies and negative viscosity appearance in the medium.

Currently, a numerical solution of the bulk of problems with allowance made for the heterogeneity of a turbulent medium, radiation, diffusion, chemistry, and MHD effects is hardly possible and only limited approaches are feasible. Note that because terrestrial planets form close to the Sun, the focus in modeling is specially narrowed to the poorly resolvable inner disk regions within several astronomical units, where matter actively accretes onto the young star. This results in the dust/gas ratio, optical opacity, and the thermal regime changes, as well as the significant contribution of photochemical processes in transformation of the matter composition and transfer.

Thermal Regime

Disc thermal regime is addressed as a key of its original inner structure and first solids formation. The viscous disc evolution with radial temperature lapse rate is regarded as the most realistic scenario of the solar system bodies' formation.

A hot rather than cold model of planets accumulation from the matter similar to chondriticmeteorites seems most plausible. It is supported by the established differences in the abundances of many elements and their isotopic ratios between the Sun, undifferentiated meteorites, and Earth. Indeed, all chondrites (except CI) and our planet are depleted in moderately volatile (Na, K, Rb, Sn, etc.) and highly volatile (Cs, Pb, etc.) elements relative to the solar abundances that are similar to their abundances in carbonaceous CI chondrites. The depletion is most pronounced in such elements as Bi, Cd, Cs, Hg, In, Pb, Se, Te, Tl, Zn, S, etc. that was found to be typical not only for various types of chondrites but also for the bulk composition of the terrestrial planets and some large planetesimals, for example, for the parent bodies of basaltic achondrites eucrites. Hence one may conclude that the differentiation of moderately and highly volatile elements was an important large-scale process at the early evolutionary stages of the protoplanetary accreting disk under relatively high temperature.

Basically, the observed depletion of highly and moderately volatile elements could result from either partial evaporation or incomplete condensation of the proto-matter of planets and the parent bodies of chondrites, because the higher the volatility of an element, the greater the depletion. Thermodynamic calculations showed that temperatures of no lower than 1200–900 K are required in both cases. Under such temperatures only non-equilibrium partial evaporation may occur and indeed, some CAIs show clear chemical and isotopic evidence of that. It is of interest to note that an experimental study of the fractional evaporation of the material of CI chondrites aimed at obtaining the material of chondrites of other types showed that, irrespective of the redox evaporation conditions, the residue obtained through heating differs radically from the actual chondrites material in the abundances of highly and moderately volatile elements. These differences were claimed to be particularly true for the abundances of such pairs of elements as Zn and Se; Sn and Pb; Rb and Cs; etc., with similar degrees of depletion relative to the abundances in chondrites. The results, however, are not unambiguous because all chondritic meteorites are cosmic sediments; that is, their components have formed at very different conditions within the nebula and experienced different thermal history before being accreted together later.

The condensation mechanism for the differentiation of moderately and highly volatile elements accompanied by their incomplete accumulation in the proto-matter of planets and the parent bodies of chondrites seems more justified. For most elements, there is a clear correlation between the degree of depletion and the temperature of its condensation from a solar-composition gas. It is most likely attributable to the influence of the kinetic constraints on the heterogeneous reactions in a gas-solid system associated with the reduction of the reaction surface of small dust particles during their accumulation and some other factors. This allowed us to conclude that much of the chondrites material formed through the condensation of the gas phase of the protoplanetary disk.

Temperature initially increased at the disk formation stage, reaching the evaporation temperature of magnesian silicates and metallic iron at a distance of 1 AU (1400 K) and then decreased in the course of viscous disk evolution. In the formation region of the terrestrial planets, it has always remained well above the H_2O condensation temperature it did not dropped below 300–500 K.

Apart from mineral phases containing moderately volatile elements, there are crystalline Mg- and Fe-bearing silicates in chondrites that support the hot protoplanetary disk formation model. One may assume that these amorphous and crystalline silicates formed as a result of the condensation processes in the inner zone of the circumstellar disks. Amorphous silicates of the interstellar molecular cloud from the fragment of which the solar system formed are believed to have evaporated in its inner zone at the early evolutionary stages to be condensed as the nano-to-micrometer sized crystalline silicate grains during subsequent cooling of the disk gaseous phase. One may suggest that the inclusions enriched in the most refractory elements (such as Ca and Al composing CAIs, as well as the rare Hf, Sc, Lu, etc.) were the earliest condensates formed near the Sun ($r < 0.5$ AU) at $T \sim 2000–1700$ K and partially carried outward through the radial drift up to the formation zone of the parent bodies of chondrites with refractory chondrules embedded in their matrix and entered rocky planetesimals. Obviously, a small fraction of them reached $r \sim 5–10$ AU, as evidenced by the findings of olivine crystals in the material of Halley and Hale-Bopp comets as well as in the Comet 9 P/Tempel nucleus investigated in the Deep Impact experiment.

Several important formation stages of the gas-dust disk around the proto-Sun and its subsequent dynamical, thermal, and cosmochemical evolution, including the non-equilibrium partial evaporation, condensation, and compaction of protoplanetary matter were traced in the framework of quite comprehensive self-consistent numerical model. In particular, it yielded the distributions of thermodynamic parameters (TP) in the viscous disk around the young Sun passing through the *T Tauri* stage, as a function of radial-vertical distance (the *rz* coordinates). Radial temperature profiles in the disk's mid-plane at the different stages of its evolution were derived, as it is shown in figure.

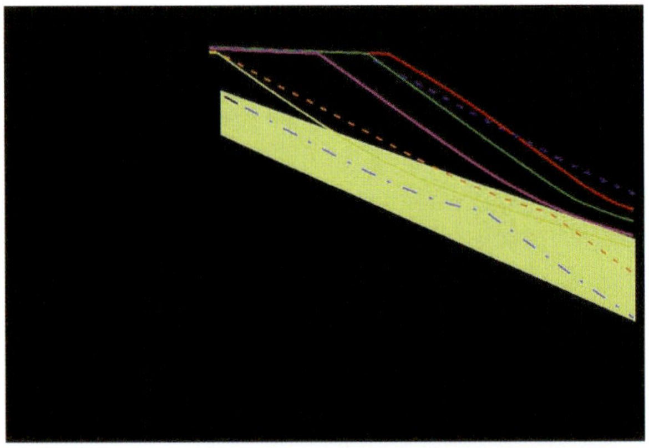

Radial temperature distribution T*(r)* in the midplane of a protoplanetary disk between 10 and 0.5 AU. 1–4 formation stage between 0.12 and 0.45 Ma (measured in millions of years) from the disk origin; 5, 6 viscous dissipation stage between 0.65 и 1.8 Ma; 7a–7в stages of dust subdisk compaction and dust clusters formation at about 2 Ma

Temperatures are controlled by the processes of condensation/evaporation of forsterite, enstatite, and iron; maximum values $T_{si} \geq 1{,}600$ K at r = 0,1 AU and $T_{si} \leq 1{,}400$ K at r = 1,4 AU. Water ice is formed at $T_w \sim 140–160$ K under gas pressure $P = 10^{-7}–10^{-4}$ bar at about 3 AU (break in the dot-dashed curve). The green zone shows the range of possible temperatures at the stage of dust clusters formation.

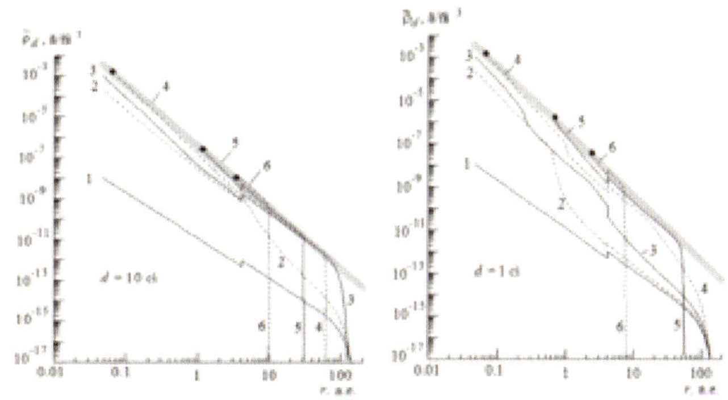

Figure: Variation of subdisk dust density ρ_d depending on the radial distance r from 100 AU for particles of mean size 10 cm (left) and 1 cm (right). Broad diagonal band critical dust density ρ_{dcr} corresponding to the disk gravity instability and its fragmentation into dust clusters. Curves 1–6 – time sequence 0, 1×10^3, 5×10^3, 2×10^4, 5×10^4 and 1×10^5 (yr) from the beginning of subdisk evolution

Variation of subdisk dust surface density ρ_d as a function of the radial distance r, the dust/gas densities ratio and size of dust particles is shown in figure as a sequence of subdisk dust density evolution, before critical surface density is achieved and the subdisc break up into dust clusters.

Dynamics

It should be specially noticed that solar system formation was intrinsically related to the dynamical processes at all stages of origin and evolution including different kind of resonances defining the system's structure, configuration, and stability. Complex nonlinear dynamical processes impose certain constraints on the models of disc evolution and planetary embryos origin. The role of dynamics is manifested, in particular, by observations of the formation of some peculiar configurations in the extrasolar planetary systems around the parent solar-type stars.

The basic principles of the dynamical systems were derived by Anri Puankare and thoroughly developed by the numerous researchers. The main parameters defining planetary system configuration are resonances and dynamical chaos. The study is used to be performed as general N-body problem or in the framework of restricted three-body approximation with involvement of some constraints. Besides resonant and chaotic motions, this approach allowed us to describe periodic orbits, collisions, escape, and other processes of the long-term dynamics based on either analytical or (mostly) on the direct numerical integration of a set of ordinary differential equations. It was shown that orbits of the solar system planets with small eccentricities and inclinations are weakly chaotic and do not exhibit essential secular variations on the time scales comparable with the solar system age. Resonances (a near-exact relation between characteristic frequencies, e.g., orbital periods, of two bodies) and tidal oscillations are pronounced in the systems of small bodies and their tidal interactions with planets, which exerts strong influence on stability of their orbits being responsible for migration processes. The stability problems including interaction of nonlinear resonances in Hamiltonian systems were analyzed in detail in the framework of Kolmogorov-Arnold-Moser (*KAM*-theory). Recently, these and some other studies laid the basis for investigation of the structure and dynamics of the systems of exoplanets around binary and multiple stars.

Migration involving both forming planets and small bodies could play a major role in the history of early solar system affecting its configuration and planets composition. Great diversity of bodies revealed by compositional mapping of the main asteroid belt implies substantial mixing of bodies of different taxonomic classes through the various dynamical processes affected by planetary migration. The outer planets' instability could cause massive influx of bodies to the inner solar system. Hydrated C-type asteroids and comets transported from outer regions of the solar system would contribute to the above diversity and also be responsible for delivery of water and volatiles to the Earth and terrestrial planets deposited as a veneer at the final stage of these planets' accumulation.

Unlike classical models of the terrestrial planets evolution with Jupiter–Saturn's stable orbit positions, there were different scenarios developed in order to explain inner solar system's water under action of the outer planets' dynamical instability. Migration of bodies and dust from the disturbed trans-Neptunian regions in Jupiter Crossing Orbits (JCO) and then in the main asteroid belt to eventually become NEOs was studied in the numerical models. The mechanism involving unstable positions of the giant planets in the course of their formation was explored by several teams. Inward migration of Jupiter from ~ 20 to ~ 5AU and getting it captured in 2:3 or 1:2 resonance with growing Saturn (which could halt and reverse the inward migration of Jupiter) was considered in the model known as the Grand Tack scenario. It was assumed that this process would shepherd outer belts' bodies inward and exert influence on planetary embryos in the inner region. This would result in volatile implantation through planetesimals' impacts during terrestrial planets formation and storage of remnant pristine bodies in the main asteroid belt, as well as the position and coexistence of asteroids of different compositions. Moreover, such a scenario was invoked to explain depletion of water in the zone of terrestrial planets' formation in terms of a "fossilized" snow line position outside Jupiter's precursor orbit beyond the snow line at ~3 AU preventing condensation of water and ice inside and/or mechanism of water delivery related to giant impacts.

Protoplanets embedded in a gaseous disc would experience drift changing their orbital positions. The movement of Uranus and Neptune from the place of their initial formation close to the Jupiter–Saturn zone further away from the Sun was assumed to exert formation of the trans-Neptunian Kuiper belt. This scenario was justified by evaluating the time required for the accretion of Uranus and Neptune in their present-day orbits. Indeed, modeling showed that such a time would exceed the age of the solar system. The model, which is sometimes referred to as reconfiguration of giant planet orbits suggested an existence of the primordial disk of several tens of Earth masses made of comet-like objects residing just outside the initial orbits of the giant planets. The disk is thought to be scattered throughout the solar system by gravitational interactions between the giant planets leading also to their migration. It is worth noting that movement of Uranus–Neptune outward due to gravitational interactions with the planetesimals would ultimately eject them into hyperbolic orbits.

The dynamical models involving instability in the giant planets zone were associated with the dramatic event of Late Heavy Bombardment (LHB) of the Moon and terrestrial planets and the Moon crater chronology. However, timing of the instability (well before the LHB or later) underlying these models remain unclear. Most recently, instead of using the terminal cataclysm hypothesis that poorly explained the triggering flux of bodies from the outer asteroid belt, linked the source of LHB projectiles with leftovers of planetesimals ("tail end of planet accretion") from the time of outer planets' formation. In support of this idea, they brought some evidence of lunar craters' records and matching density of craters with the Highly Siderophile Elements (HSE) in the lunar mantle. These "iron loving" elements were found in chondrite meteorites and presumably they were delivered with projectiles of similar composition after a major event of the lunar matter differentiation, with the core emergence and sinking of the original HSE in the core. Hence, geochemical constraints of large craters' composition formed by projectiles and the timescale of mantle ocean crystallization (~ 4.35 Ma, with the account for Fe-S segregation), advocate for the imprint of LHB event at the tail end of planet accretion. The modern model coupling LHB at ~ 3.9 Ma with the time of outer planets' instability appears to accommodate the current knowledge with minimum assumptions.

The critical role of C-chondrites and water transport in the inner solar system could be played by the process of outer planets' formation with no change of their orbital positions. This is a kind of scattering model that postulates a slow accretion of gas in the broad outer belt rich of volatiles (~5–20 AU) by the giant's cores until some threshold with the onset of Jupiter's runaway growth. This process provides destabilization, gravitational scattering of nearby planetesimals by Jupiter and Saturn and their large-scale radial mixing, some of them moving inward and residing (resonance) orbits' interior to that of Jupiter. As gas dissipates, orbits of the disc planetesimals become progressively eccentric, including those trapped in the main asteroid belt and crossing the region of growing Earth and terrestrial planets, thus delivering water with C-chondrites and other type of meteorites. Later implantation of these species by scattering triggered by growth and possible migration of Saturn and icy planets turns out to be non-efficient. This scenario with no or very limited key assumptions related to the Grand Tack or Nice models seems quite plausible and implies a close relationship of both inner and outer planets' formation.

Particles Growth and Solid Bodies Formation

The key problem of the solar system origin is how the solar system bodies (original dust and condensates, as well as those produced in coagulation) progressively grew on scales ranging from nano- and micron-size particles to planetesimals and planets thus ranging over dozens orders of magnitude in mass. As mentioned above, time-dependent modeling with the different particle size distribution functions and limited lab experiments is the only tool to gain insight into the problem. Numerous attempts to reconstruct the process taking into account nebula thermal structure; evaporation fronts (EFs') position for different components depending on radial temperature lapse rates in the disc's midplane; particle growth by sticking limited by bouncing; fragmentation and radial drift due to nebula headwind drag of growing bodies influencing mass redistribution; etc., have been undertaken. Also, lab experiments with silicate and ice particles collisions in microgravity conditions revealed some important patterns of particles and particle aggregates formation. Some important constraints were deduced for particles bouncing and sticking, including translational energy and coefficients of restitution estimates depending on impact parameters.

According to the modern views, the process started owing to the above-mentioned streaming/gravitational instability development in the dense subdisk formed out of the dust component settled down to the midplane of the gas-dust disk. This was followed by the primary porous dust clusters formation from which the first solid bodies of pebble-boulder size and eventually planetesimals of asteroid size have emerged. Some of these cm-size and larger pebbles were possibly assembled into porous clumps giving birth to comets; this approach accommodates the primordial rubble pile theory though other theory postulates that comets were made out of debris left over from the main planet-building phase, thus preserving remnants of the protosolar nebula matter.

Nonetheless, collisional rubble pile theory better explains a rather small size of comet's nucleus (~5–10 km) and even the bi-lobed shape of some of them, because of their gradual formation at low speed of leftover pebbles/grains after larger bodies have accumulated and gas in disc has disappeared, followed by gentle collisions owing to stirring the cometary orbits, specifically at the skirt of planetary system.

The sequence of solids growth and timescale of such a scenario is shown schematically in figure below.

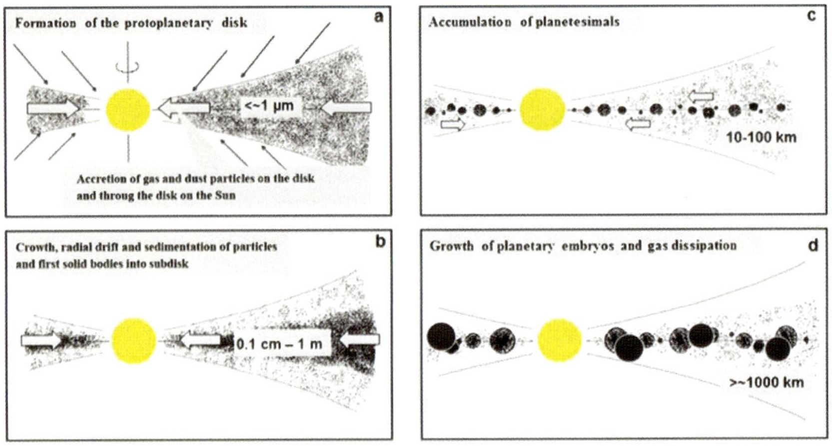

Sequence of the of protosolar disk evolution

(a) Disk formation due to accretion gas and dust from the protosolar nebula and protosun emergence in the centre. (b) Disk flatteniand dust particle sedimentation toward the midplane and dense dust subdisc formation; particle growth. (c) Subdisc fragmentation into dust clusters due to streaming and gravity instability development, cluster collisional interaction and solid growth, including first proto-planetesimal accumulation with gravity domain. (d) Planetesimals and planetary embryo formation, gas dissipation, and original solar system architecture setup evolving ultimately to the contemporary configuration. Time span: (a)–(b) 10^5–10^6 yr; (c)–(d) 10^6–10^7 yr.

The initial process is thought to last less than 10^5–10^6 years of the overall ~10^8 years of planets formation. Note that the same short timescale is assumed for comets' formation as well as for initial growth phase of the TNOs probably influencing the cometary-like bodies origin and evolution. The shear turbulence mechanism would promote ring-like contraction of dust in protoplanetary cloud into a thin disc ($h << r$) of non-regular shape widening toward the periphery. Solid bodies becoming planet embryos are formed from initially "loose" gas-dust (porous) clumps filling the main part of their sphere of attraction (Hill's sphere) and slowly contracting due to internal gravitational forces. Let us recall that whereas the growth of particles during their collisions is hampered in chaotic turbulence, their coalescence and enlargement can occur inside turbulent eddies' coherent structures promoting dust clusters set up in vortexes of baroclinic nature in a broad range of Stokes (St) number defined by the ratio of particle drag in an ambient gas to characteristic time of the system (defined as inverse Kepler frequency Ω^{-1}) and dependent on size and density of a particle. Solid particles may be also concentrated in the disks with quite weak turbulence where denser than average matter structures (with large dust-to-gas ratio) would create a large population of aggregates and trigger streaming instability. This could be the case in outer zone of massive discs where rapid growth of aggregates to planetesimals was suggested.

Generally, both instability mechanisms support the basic Safronov and Goldreich-Ward ideas about disc viscous accretion and subdisk matter gravitational collapse. As we have seen, in the modern models, the focus is also given to the competing processes of gas and dust photo-evaporation by the solar EUV-X-ray radiation and gas condensation/condesate growth, as well as to dust trapping, clustering, and coagulation including particles size distribution and stratified turbulence. These processes are accompanied by dust aggregates formation/restructuring through particles sticking, gas-dust coupling decrease, solids growth, and radial drift resulting ultimately in gravitationally bound planetesimals and planet embryos origin.

Although the basic scenario of ongoing particle growth is generally understood, many details of the processes involved at the different stages remain unclear. First, still uncertain are the details of physical mechanisms responsible for primary small dust particles growth before gravitational interactions of hundred meters-kilometer size bodies become effective. The process of particles' mutual collisions is usually invoked as the factor that presumably gave rise to the aggregation of small particles to yield either dense particle clumps or pebble/boulder-size compact aggregates migrating in the protoplanetary disc and controlling planetesimals' growth. However, a probability of destruction rather than growth of forming bodies may be higher if collisional relative velocities and bouncing barriers for particle aggregates are significant, until gravitational attraction of planetesimals becomes dominant. Hence, an efficiency of sticking of dust particles into aggregates through collisions is a rather delicate mechanism, specifically if equal size particles of ~mm size are considered. The formation of larger objects limited by bouncing is thought to be augmented by the streaming instability, gravitational collapse, collective particle behavior, and/or by static compression of fluffy dust aggregates.

As a compromise ensuring higher efficiency of agglomerates creation we advocate for collisional integration of fluffy clusters made up by micron-sized particles rather than individual particles themselves, in support of the lab experiments mentioned above. It is further assumed in the developed models that clusters and their individual particles have fractal structure. Such an approach was thoroughly explored and validated theoretically involving concurrence of gravitational and brownian coagulation of dust monomers, aggregates growth, and porous bodies interaction/growth.

The bottom line of our numerical model is the agglomeration of small particles in the primary fluffy dust aggregates of low-bulk density and fractal nature of the latter, giving rise to their compression and formation of first dense bodies of progressively larger size. Unlike an approach used in the previous simulations, the mode of particle on-head and offset collisions inside dust clusters was more strictly assessed in numerical N-body models. Method of permeable particles and modified Newton model of collisions were utilized in terms of restitution coefficient Kr dependent on the distance between centres of particles and their relative velocity, taking into account internal structure of particles in fluffy clusters and complicated patterns of their interactions, specifically in the contact zone. This enabled us to examine collisional evolution of fluffy clusters beginning from nano-particles sticking together through electrostatic forces and growing up to compressed aggregates of larger size and different patterns. Example of 3D modeling of evolution of fractal dusty clusters with different numbers of populated submicron particles is shown in figures.

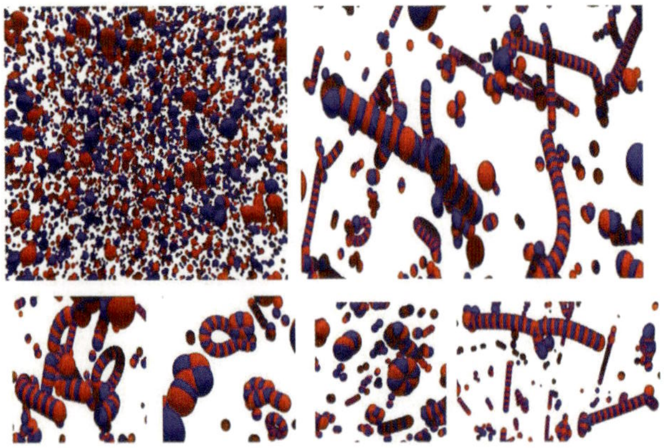

*Figure:*3D-numerical modeling of submicron particles, collisional interaction, and growing within a grid in the primordial dusty cloud (density $d = 1.5 \times 10^{-21}$ g/nm^3. (a) An example of the original 3D structure of a selected grid. (b) Examples of agglomerate formation in the process of collisional evolution of particles in some grids

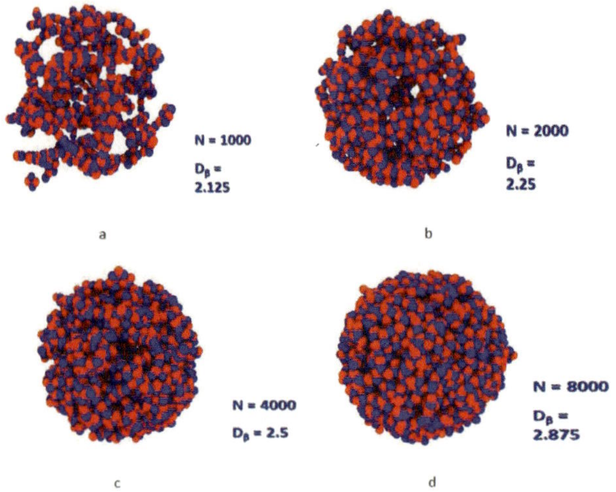

Figure: 3D structure of fractal dusty clusters with different numbers of populated particles evolution (a–d) Red and blue colors denote positively and negatively charged particles of about 20-nm characteristic size in a quasi-neutral medium

Figure is a snapshot of a cross-section of randomly selected grids for fractal dusty clusters with different numbers of populated particles in figure.

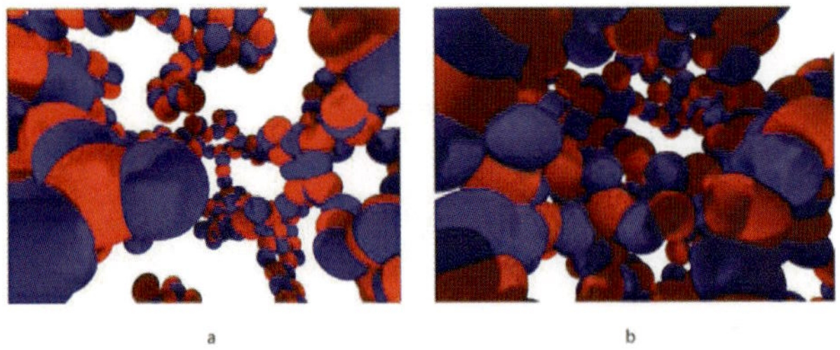

Figure: Snapshot of a cross section of a randomly selected grid when getting through fractal dusty clusters with different numbers of populated particles (clusters a and b, respectively)

Of special interest is the evolution of fluffy dust clusters in mutual collisions that is critical for dusty agglomerates formation, as shown in figure.

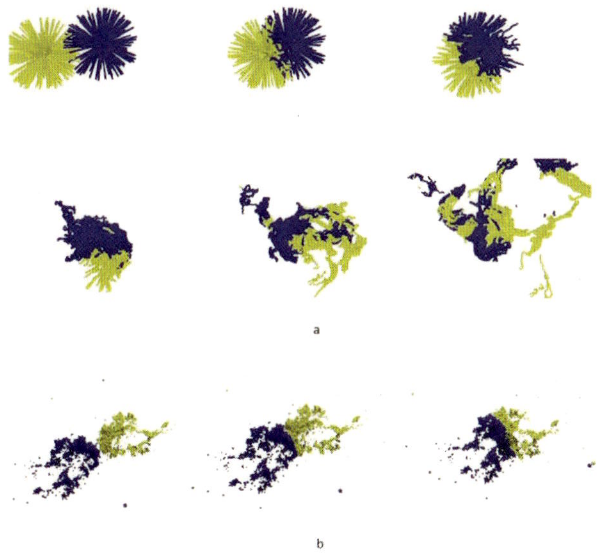

Figure: Evolution of the structures of fluffy dust clusters; computer modeling of collisional interactions. (a) Fluffy clusters of characteristic size 50 nm, fractal dimension 2.55, and number of particles = 8,192. (b) Fluffy clusters of characteristic size 75 nm, fractal dimension 2.15, and number of particles = 3,072

Examples of the respective numerical experiments carried out with clusters composed of various number of submicron particles of different fractal dimensions D_β ranging from 2.025 (very fluffy structure) to 2.975 (well-packed structure) are shown in figure. The results of modeling of cluster interactions reveals how K_r (restitution coefficient or its equivalent collision recovery ratio) changes for clusters of different D_β depending on the relative distance between colliding *i-j* particles (0–1), and collision energy E_{ij}/E_{int} (E_{ij} is kinetic energy of collision; E_{int} the internal energy of molecular coupling preserving structure) ranging from 0.01 to 0.99. The strong dependence of K_r on the parameters involved and are bound threshold were found, for example, for dense colliding clusters with relative velocity less than the critical value(corresponding to the energy of plastic deformation)bouncing occurs at $K_r > 0.87$.

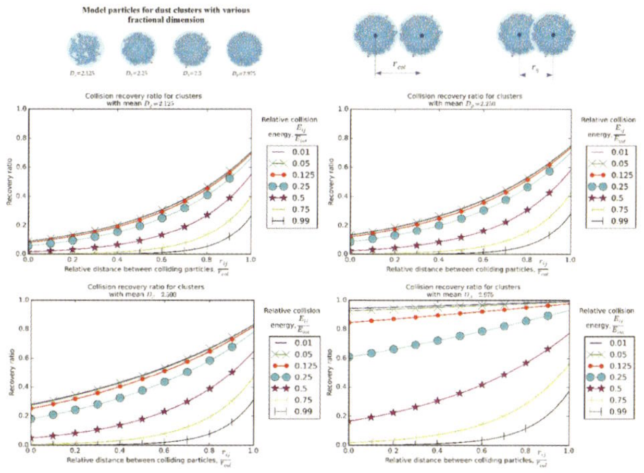

Figure: Restitution coefficient K_r (collision recovery ratio) for clusters with various fractal dimensions $D\beta$ as shown here. Note: r_i, r_j — radii of *i* and *j* model particles; r_{ij} — distance between *i* and *j* colliding model particles; $r_{col} = r_i + r_j$ — collision distance; $\frac{r_{ij}}{r_{col}}$ — relative distance between *i* and *j* colliding particles; *E int*: cluster's internal energy; *E ij*: kinetic energy of collision; $\frac{E_{ij}}{E_{int}}$ — relative collision energy

The results are regarded as an important milestone toward in-depth study of "cluster-cluster coagulation" at the initial stage of the protoplanetary gas-dust disc evolution and results are addressed as precursors of primary objects growing. They are in accord with the conclusion that a cluster with a larger number of particles is harder to destroy (easier to survive) in energetic collisions. Nonetheless, the problem is still far from being solved, especially if one keeps in mind that the global qualitative effect of disk gravity further increases collision/impact velocities and adds additional jitter to the orbital evolution of the primary bodies.

The next stage of growth of primordial seeds of bodies to planetesimals, with the allowance for mechanism of self-gravitation, seems more conceivable. At the first approximation, the mass distribution of particle agglomerates and proto-planetesimals obeys the known Smoluhovsky coagulation equation (a sort of the Boltzmann kinetic equation when dealing with coagulation process) with the account for gravity mutual attraction and fragmentation in non-gentle collisions. This interaction along with the dust sedimentation onto the formed bodies residing on intersected orbits (with chaotic velocities superimposed on quasi-circular orbital velocities) results in further growth of these protoplanetary embryos followed by the gradual scooping up of smaller bodies in due course of the swarm evolution. It is supposed, however, that there is a sort of barrier at around one meter size-range because of the effects of body destruction and their inward drift toward the central star. The latter is caused by aerodynamic braking of these bodies in the remaining gas whose rotation velocity becomes lower than the Keplerian one. Primordial bodies beyond the meter-size barrier to proto-planetesimals growth through "coagulation" after more complete gas evacuation seems more feasible.

Obviously, competitive processes of destruction and particle trapping during radial matter exchange, along with particle concentration via streaming and/or gravitational instability, eventually results in bodies' growth. Based on the results of modeling supported by lab simulation millimeter- to centimeter-size pebbles in a gravitationally bound collapsed cloud (due to, for example, disk streaming instability) would experience numerous collisions with different outcomes depending on their relative speeds and the cloud density, resulting eventually in a quite intense formation of planetesimals. As far as a planet's accumulation is concerned, gravitationally focused mutual collisions of asteroids and planetary embryos are thought to be responsible for the emergence of progressively growing proto-planets possessing larger gravitational potentials (oligarchs), such an oligarchic growth being accompanied by catastrophic fragmentation of other bodies in a swarm of planetesimals. A system of satellites around a proto-planet could be explained invoking the same mechanism.

A possibility of the formation of bodies from dust clusters in mass range from 10^{20} to 10^{22} g (corresponding to asteroids of tens to hundred kilometers in size) was shown in the numerical model. A mechanism of gravitational collapse of massive swarms of small particles and their pile-up without passing through the intermediate phases of growing was proposed to explain early formation of the ~100 km bodies, a source of very old iron meteorites, thereby circumventing the meter-size barrier problem. Similarly, a prompt planetesimal's formation in regions beyond the snow line due to streaming instability-induced gravity collapse of radially drifting millimeter-size dust particles to ensure material deposition for the giant planet cores was suggested. Anyway, whatever particular solution of the problem, further growth occurs through interaction of planetesimals with solid and gaseous components under gravity domain and sporadic mutual collisions.

Note that in the numerical experiments, some constraints were placed on the value of angular momentum of the porous (low density) gas-dust clusters (pre-planetesimals) of about Hill sphere in size and their moving before collision along heliocentric orbits. The model aimed to assist dynamical formation of planetary/satellite bodies with application to trans-Neptunian system. Arguments were drawn for consistency of the properties of wide binaries in the Kuiper Belt with a primordial origin during gravitational collapse. As the outcome, origin of the Earth-Moon system was attempted to explain in terms of the compression of a low-density cluster of 0.1 mass of the contemporary Earth formed from the collision of two original clusters on the close heliocentric orbits, which allowed the system to acquire the necessary angular momentum.

As it was said above, significant role at the final stage of planet's growth could play migration and resonances in the forming planetary system. The commonly adopted scenario of the solar system origin involving planetesimals growing to planet embryos together with asteroid-size and comet-like bodies as remnants of planets is illustrated in figure.

Artist's concept of a planetary system formation, as seen from the outer edge of the gas-dust disc around an emerged central protostar. Planetesimals growing to planet embryos, together with asteroid-size and cometlike bodies as remnants of planets' formation, are seen. The picture resembles the commonly adopted scenario of the solar system origins

It reflects our current general understanding of how original bodies and then planets formed though we are still far from resolving this mysterious subject..

Inner Planets Evolution

The key question intrinsically related to the origin of the terrestrial planets is why the three neighbour planets Earth, Venus, and Mars inheriting disc composition in a close neighbourhood and basically, exhibiting similar chemistry in the formation took different evolutionary paths? Namely, what caused, unlike the pleasant climate of Earth, hostile thermal regime of Venus, and why the assumed original clement climate on Mars has drastically changed?

To answer this question, many complicated problems closely related to planetary cosmogony should be addressed. Some of them relevant to this important topic were considered in previous chapters of the paper and they are discussed in more detail in some reviews and monographs. Here, we focus only on limited aspects of the problems involved that concerns a rather simple approach in the treatment of thermal regime formation on the inner planets and the fate of water as the key climatic factor impacting on their evolution.

Let us first note that Earth is located in a comparatively narrow zone of circumsolar space where the development of favourable (for life's existence) climatic conditions is possible. This is the so-called habitable zone in the solar system whose inner boundary is located 10–15 million km closer to the Sun than the Earth's orbit while its outer boundary extends to approximately the orbit of Mars. The orbit of Venus turns out to be outside this zone, at a distance that is almost triple the critical value. Obviously, if the Earth were moved to the place of Venus (actually, even less) then it would probably evolve according to the Venusians' scenario.

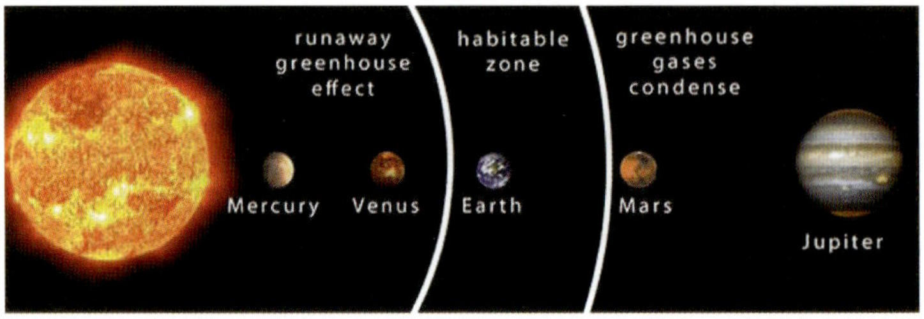

Figure: The habitable zone in the vicinity of Earth

The zone is limited by only about 10–15 million km toward the Sun and extends outward to nearly the Mars orbit. The planet shifting closer to the Sun would meet runaway greenhouse similar to that on Venus (T_s = 735 K; P_s = 92 atm), whereas at Mars's position, greenhouse gases would condense out and the atmosphere would collapse. Unlike Earth, the atmospheres of Venus and Mars consist mainly of carbon dioxide; their abundance is controlled by the carbonate-vollastonite equilibrium between atmosphere and crust, which is strongly dependent on surface temperature T_s . CO_2 storage in the Venus atmosphere is the same as what is locked in the Earth's crust, and it would release into the atmosphere amounting to $P_s \sim 90$ atm if the clement climate on the home planet irreversibly changed (temperature dramatically raised) and the CO_2 balance maintenance ceased.

Based on the average annual balance of the solar radiation absorbed by the terrestrial surface and atmosphere with an account of its albedo the effective (equilibrium) temperature of Earth is $T_e \sim 255$ K, which is even below the freezing point of salt water. This means that favourable climate conditions could hardly have arose on the home planet unless the greenhouse effect caused by absorption of solar radiation in its relatively thick atmosphere would raise the mean temperature near the surface to the pleasant 288 K. This atmosphere of secondary origin, unlike the primary atmosphere captured from the protosolar nebula and composed of light atmophile elements that were lost, formed owing to degassing of volatiles from the interior and from collisions with remnant planetesimals and migrating bodies. It presumably contained carbon dioxide, nitrogen, water vapor, and some minor species including such greenhouse gas like methane.

Earth could therefore retain its water on the surface, which was mainly concentrated in the oceans. In turn, carbon dioxide would accumulate incarbonates of sedimentary rocks through its binding with metal oxides incorporated into the ocean crust and upper mantle (with the formation of aqueous silicates) and biogenically, through the deposits of lime skeletons of sea organisms. Dramatic change of the Earth's atmosphere with its modern composition occurred much later, about 2.8 Ma, due to bacterial production of oxygen accumulated in the planet's gas envelope, which made possible the life forms rich diversity and proliferation.

In contrast, equilibrium temperature of the early Venus is supposed to be not lower than $T_e \sim 325$ K, which is above the water boiling temperature in a low pressure environment. One may reasonably expect surface pressure on ancient Venus as an Earth-like plane to be substantially lower than 1 bar. Under these conditions, carbon dioxide would gradually accumulate in the atmosphere together with water vapors, which in turn, have contributed to a further rise in surface temperature. The runaway greenhouse eventually developed resulting in the transport of increasingly large amounts of CO_2 and H_2O into the atmosphere, up to some equilibrium state corresponding temperature 735 K and pressure 92 atm. At such a temperature, carbon dioxide turned out to be not bound in carbonates of sedimentary rocks, as on Earth but was released into the atmosphere giving rise to very high pressure. According to the model estimates, the amount of carbon dioxide locked in the Earth's sedimentary rocks is comparable to the content of CO_2 in the Venus atmosphere and it would constitute nearly the same pressure had the Earth's temperature raised to similar temperature and CO_2 released in the atmosphere. The latter is characterized by the relationships between mineral phases and volatiles on the surface, with the carbonate-silicate interaction in the upper layer of the planetary crust being the most important. The above scenario is characteristic of a system with positive feedback and internal instability when initial perturbation is not suppressed but, on the contrary, fairly rapidly enhanced.

As far as water is concerned, its accumulation on Venus with the involvement of both endogenic and exogenic sources, likewise on Earth and Mars, would result in the initial amounts of water nearly equal to the volume of the terrestrial hydrosphere. However, water could not be preserved on the Venus surface at the temperature above the critical point (647 K), even as salty aqueous solutions (brines) with a slightly higher critical temperature (675–700 K). As to the atmosphere, the amount of water in it is negligible. Thus, the key problems of the planet evolution are whether water was on early Venus, where it was stored, and, if so, when and how was it lost. The deuterium-to-hydrogen ratio measured in the Venusian atmosphere, which turned out to be larger than that in the Earth's atmosphere by two orders of magnitude, confirms the concept of ancient Venus ocean because high deuterium enrichment can be explained by an efficient thermal escape of the lighter hydrogen isotope and leaving behind heavier deuterium. Note that adoption of this mechanism requires an assumption that enormous masses of liberated hydrogen and oxygen through dissociation by the solar EUV radiation were removed from the atmosphere by efficient dissipation and bound by surface rocks, respectively, which seems, however, hardly realistic.

There is also a different viewpoint that Venus was initially formed as a "dry" planet, though the introduction of volatiles through the mechanism of heterogeneous accretion is assumed to play an important role as well. In this scenario, the water content was essentially constant throughout the geological history of Venus, remaining approximately at the current level while the efficiencies of heterogeneous accretion and hydrogen dissipation from the atmosphere were considerably lower. In such a case, the observed enrichment of the Venusian atmosphere in deuterium during the isotope fractionation in the process of their dissipation from the atmosphere can be also explained.

Addressing now the problem of Mars evolution, we start from a well-supported concept that the quite pleasant natural conditions presumably were on Mars at early epoch, which is backed by the morphological features preserved on the planet's surface as well as by some other evidence. Since equilibrium temperature of early Mars would not exceed T_e~220 K it would be able to retain only frozen water on its surface. However, there is widespread evidence of the presence of liquid water flows meaning considerably denser atmosphere and much more favourable climate ~3.8–3.5 billion years ago. Obviously a catastrophic drought suddenly occurred on Mars rather than a long-term evolution has resulted in the present-day natural conditions. This dramatic change cannot be explained by the fact that the planet is 0.5 AU farther away from the Sun than Earth. More likely it should be associated with the fact that the mass of Mars is almost an order of magnitude lower than that of the Earth implying a lower storage and an early depletion of the former in the long-lived radiogenic isotopes, which served as the main energy sources responsible for the thermal history of a rocky planet. Mars began to cool, volcanism ceased, atmosphere diminished. In other words, a limited inventory of radiogenic isotopes rather than the distance from the Sun has dramatically impacted the Mars evolution changing its geology and causing collapse of the atmosphere. Likewise Earth and Venus, Mars probably possessed an ocean, currently its significant fraction being preserved beneath the surface as underground water.

It is important to emphasize that synergy of the Earth and planetary sciences is aimed at a better understanding of the present, past, and future of the Earth based on comparative planetology approach. It is intended to answer the key questions about feedback for the existing regulation mechanisms to maintain clement climate. Concurrently, this contributes to the cardinal problems of planetary cosmogony allowing us to put more stringent constraints on the range of parameters used in modeling of the origin and evolution of the solar and other planetary systems.

Ring System

Planetary rings are not solid objects. They are rings of dust and other particles that are in orbit around a planet- almost like billions upon billions of tiny moons all orbiting close together so that from a distance the rings actually look like a hoop around a planet.

The most famous planetary rings are without a doubt those of Saturn, but the other gas giants in our solar system (Jupiter, Uranus and Neptune) also possess less impressive ring systems of their own. In 2008 the Cassini spacecraft found evidence of the possibility around Rhea, one of Saturn's moons. If this evidence is corroborated, Rhea will be the only body other than a planet to have a ring system.

What Causes Rings?

A ring is a collection of vast numbers of particles, each like a tiny moon obeying Kepler's laws as it follows its own orbit around the planet. Thus, the inner particles revolve faster than those farther out, and the ring as a whole does not rotate as a solid body. In fact, it is better not to think of a ring rotating at all, but rather to consider the revolution (or motion in orbit) of its individual moonlets.

If the ring particles were widely spaced, they would move independently, like separate moonlets. However, in the main rings of Saturn and Uranus the particles are close enough to exert mutual gravitational influence, and occasionally even to rub together or bounce off each other in low-speed collisions. Because of these interactions, we see phenomena such as waves that move across the rings just the way water waves move over the surface of the ocean.

There are two basic ideas of how such rings come to be. First is the breakup hypothesis, which suggests that the rings are the remains of a shattered moon. A passing comet or asteroid might have collided with the moon, breaking it into pieces. Tidal forces then pulled the fragments apart, and they dispersed into a disk. The second hypothesis, which takes the reverse perspective, suggests that the rings are made of particles that were unable to come together to form a moon in the first place.

In either theory, the gravity of the planet plays an important role. Close to the planet, tidal forces can tear bodies apart or inhibit loose particles from coming together. We do not know which explanation holds for any given ring, although many scientists have concluded that at least a few of the rings are relatively young and must therefore be the result of breakup.

Four Ring Systems

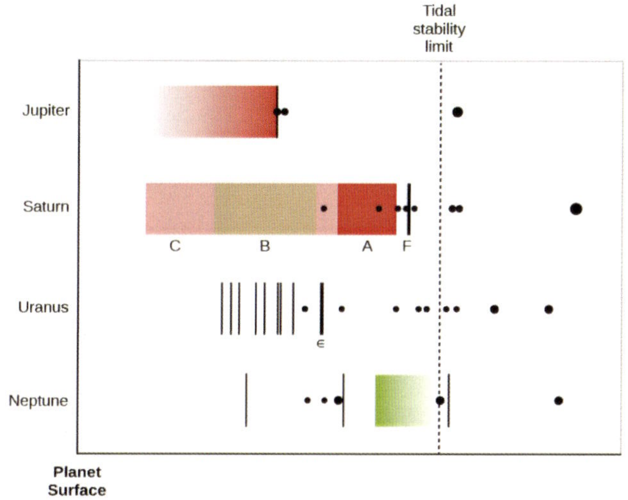

This diagram shows the locations of the ring systems of the four giant planets

The left axis represents the planet's surface. The dotted vertical line is the limit inside which gravitational forces can break up moons (each planet's system is drawn to a different scale, so that this stability limit lines up for all four of them). The black dots are the inner moons of each planet on the same scale as its rings. Notice that only really small moons survive inside the stability limit.

Rings of Saturn

Saturn's rings are one of the most beautiful sights in the solar system. From outer to inner, the three brightest rings are labeled with the extremely unromantic names of A, B, and C Rings. Table gives the dimensions of the rings in both kilometers and units of the radius of Saturn, R_{Saturn}. The B Ring is the brightest and has the most closely packed particles, whereas the A and C Rings are translucent.

The total mass of the B Ring, which is probably close to the mass of the entire ring system, is about equal to that of an icy moon 250 kilometers in diameter (suggesting that the ring could have originated in the breakup of such a moon). Between the A and B Rings is a wide gap named the Cassini Division after Gian Domenico Cassini, who first glimpsed it through a telescope in 1675 and whose name planetary scientists have also given to the Cassini spacecraft exploring the Saturn system.

Saturn's Rings as Seen from Above and Below.

(a) (b)

(a) The view from above is illuminated by direct sunlight. (b) The illumination seen from below is sunlight that has diffused through gaps in the rings

Selected Features in the Rings of Saturn			
Ring Name[1]	**Outer Edge (R_{Saturn})**	**Outer Edge (km)**	**Width (km)**
F	2.324	140,180	90
A	2.267	136,780	14,600
Cassini Division	2.025	122,170	4590
B	1.949	117,580	25,580
C	1.525	92,000	17,490

Saturn's rings are very broad and very thin. The width of the main rings is 70,000 kilometers, yet their average thickness is only 20 meters. If we made a scale model of the rings out of paper, we would have to make them 1 kilometer across. On this scale, Saturn itself would loom as high as an 80-story building. The ring particles are composed primarily of water ice, and they range from grains the size of sand up to house-sized boulders. An insider's view of the rings would probably resemble a bright cloud of floating snowflakes and hailstones, with a few snowballs and larger objects, many of them loose aggregates of smaller particles.

Artist's Idealized Impression of the Rings of Saturn as Seen from the Inside.

Note that the rings are mostly made of pieces of water ice of different sizes. At the end of its mission, the Cassini spacecraft is planning to cut through one of the gaps in Saturn's rings, but it won't get this close.

In addition to the broad A, B, and C Rings, Saturn has a handful of very narrow rings no more than 100 kilometers wide. The most substantial of these, which lies just outside the A Ring, is called the F Ring; its surprising appearance is discussed below. In general, Saturn's narrow rings resemble the rings of Uranus and Neptune.

There is also a very faint, tenuous ring, called the E Ring, associated with Saturn's small icy moon Enceladus. The particles in the E Ring are very small and composed of water ice. Since such a tenuous cloud of ice crystals will tend to dissipate, the ongoing existence of the E Ring strongly suggests that it is being continually replenished by a source at Enceladus. This icy moon is very small only 500 kilometers in diameter but the Voyager images showed that the craters on about half of its surface have been erased, indicating geological activity sometime in the past few million years. It was with great anticipation that the Cassini scientists maneuvered the spacecraft orbit to allow multiple close flybys of Enceladus starting in 2005.

Those awaiting the Cassini flyby results were not disappointed. High-resolution images showed long, dark stripes of smooth ground near its south pole, which were soon nicknamed "tiger stripes". Infrared measurements revealed that these tiger stripes are warmer than their surroundings. Best of all, dozens of cryovolcanic vents on the tiger stripes were seen to be erupting geysers of salty water and ice. Estimates suggested that 200 kilograms of material were shooting into space each second not a lot, but enough for the spacecraft to sample.

Enceladus

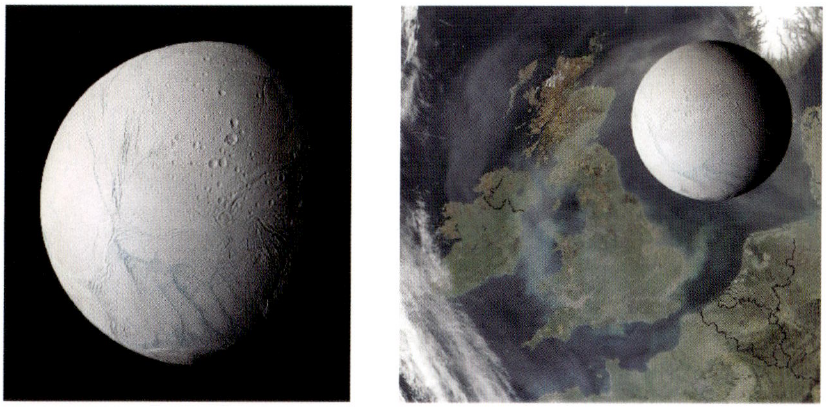

(a) (b)

(a) This image shows both smooth and cratered terrain on Saturn's moon, and also "tiger stripes" in the south polar region (lower part of image). (b) Here Enceladus is shown to scale with Great Britain and the coast of Western Europe, to emphasize that it is a small moon, only about 500 kilometers in diameter

These dark stripes (shown here in exaggerated color) have elevated temperatures and are the source of the many geysers discovered on Enceladus. They are about 130 kilometers long and 40 kilometers apart.

When Cassini was directed to fly into the plumes, it measured their composition and found them to be similar to material we see liberated from comets. The vapor and ice plumes consisted mostly of water, but with trace amounts of nitrogen, ammonia, methane, and other hydrocarbons. Minerals found in the geysers in trace amounts included ordinary salt, meaning that the geyser plumes were high-pressure sprays of salt water.

Based on the continuing study of Enceladus' bulk properties and the ongoing geysers, in 2015 the Cassini mission scientists tentatively identified a subsurface ocean of water feeding the geysers. These discoveries suggested that in spite of its small size, Enceladus should be added to the list of worlds that we would like to explore for possible life. Since its subsurface ocean is conveniently escaping into space, it might be much easier to sample than the ocean of Europa, which is deeply buried below its thick crust of ice.

Geysers on Enceladus

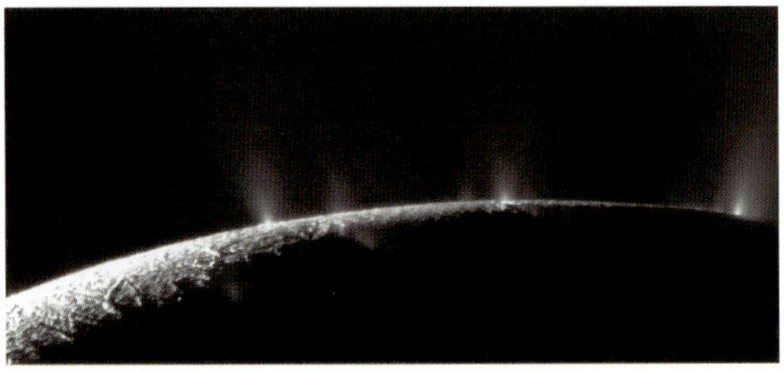

This Cassini image shows a number of water geysers on Saturn's small moon Enceladus, apparently salty water from a subsurface source escaping through cracks in the surface. You can see curved lines of geysers along the four "tiger stripes" on the surface

Rings of Uranus and Neptune

Uranus' rings are narrow and black, making them almost invisible from Earth. The nine main rings were discovered in 1977 from observations made of a star as Uranus passed in front of it. We call such a passage of one astronomical object in front of another an occultation. During the 1977 occultation, astronomers expected the star's light to disappear as the planet moved across it. But in addition, the star dimmed briefly several times before Uranus reached it, as each narrow ring passed between the star and the telescope. Thus, the rings were mapped out in detail even though they could not be seen or photographed directly, like counting the number of cars in a train at night by watching the blinking of a light as the cars successively pass in front of it. When Voyager approached Uranus in 1986, it was able to study the rings at close range; the spacecraft also photographed two new rings.

Rings of Uranus

The Voyager team had to expose this image for a long time to get a glimpse of Uranus' narrow dark rings. You can see the grainy structure of "noise" in the electronics of the camera in the picture background

The outermost and most massive of Uranus' rings is called the Epsilon Ring. It is only about 100 kilometers wide and probably no more than 100 meters thick (similar to the F Ring of Saturn). The Epsilon Ring encircles Uranus at a distance of 51,000 kilometers, about twice the radius of Uranus. This ring probably contains as much mass as all of Uranus' other ten rings combined; most of them are narrow ribbons less than 10 kilometers wide, just the reverse of the broad rings of Saturn.

The individual particles in the uranian rings are nearly as black as lumps of coal. While astronomers do not understand the composition of this material in detail, it seems to consist in large part of carbon and hydrocarbon compounds. Organic material of this sort is rather common in the outer solar system. Many of the asteroids and comets are also composed of dark, tarlike materials. In the case of Uranus, its ten small inner moons have a similar composition, suggesting that one or more moons might have broken up to make the rings.

Neptune's rings are generally similar to those of Uranus but even more tenuous. There are only four of them, and the particles are not uniformly distributed along their lengths. Because these rings are so difficult to investigate from Earth, it will probably be a long time before we understand them very well.

Rings of Neptune

This long exposure of Neptune's rings was photographed by Voyager 2.
Note the two denser regions of the outer ring

Interactions between Rings and Moons

Much of our fascination with planetary rings is a result of their intricate structures, most of which owe their existence to the gravitational effect of moons, without which the rings would be flat and featureless. Indeed, it is becoming clear that without moons there would probably be no rings at all because, left to themselves, thin disks of small particles gradually spread and dissipate.

Most of the gaps in Saturn's rings, and also the location of the outer edge of the A Ring, result from gravitational resonances with small inner moons. A resonance takes place when two objects have orbital periods that are exact ratios of each other, such as 1:2 or 2:3. For example, any particle in the gap at the inner side of the Cassini Division of Saturn's rings would have a period equal to one-half that of Saturn's moon Mimas. Such a particle would be nearest Mimas in the same part of its orbit every second revolution. The repeated gravitational tugs of Mimas, acting always in the same direction, would perturb it, forcing it into a new orbit outside the gap. In this way, the Cassini Division became depleted of ring material over long periods of time.

The Cassini mission revealed a great deal of fine structure in Saturn's rings. Unlike the earlier Voyager flybys, Cassini was able to observe the rings for more than a decade, revealing a remarkable range of changes, on time scales from a few minutes to several years. Many of the features newly seen in Cassini data indicated the presence of condensations or small moons only a few tens of meters across imbedded in the rings. As each small moon moves, it produces waves in the surrounding ring material like the wake left by a moving ship. Even when the moon is too small to be resolved, its characteristic waves could be photographed by Cassini.

One of the most interesting rings of Saturn is the narrow F Ring, which contains several apparent ringlets within its 90-kilometer width. In places, the F Ring breaks up into two or three parallel strands that sometimes show bends or kinks. Most of the rings of Uranus and Neptune are also narrow ribbons like the F Ring of Saturn. Clearly, the gravity of some objects must be keeping the particles in these thin rings from spreading out.

As we have seen, the largest features in the rings of Saturn are produced by gravitational resonances with the inner moons, while much of the fine structure is caused by smaller embedded moons. In the case of Saturn's F Ring, close-up images revealed that it is bounded by the orbits of two moons, called Pandora and Prometheus. These two small moons (each about 100 kilometers in diameter) are referred to as shepherd moons, since their gravitation serves to "shepherd" the ring particles and keep them confined to a narrow ribbon. A similar situation applies to the Epsilon Ring of Uranus, which is shepherded by the moons Cordelia and Ophelia. These two shepherds, each about 50 kilometers in diameter, orbit about 2000 kilometers inside and outside the ring.

Saturn's F Ring and its Shepherd Moons

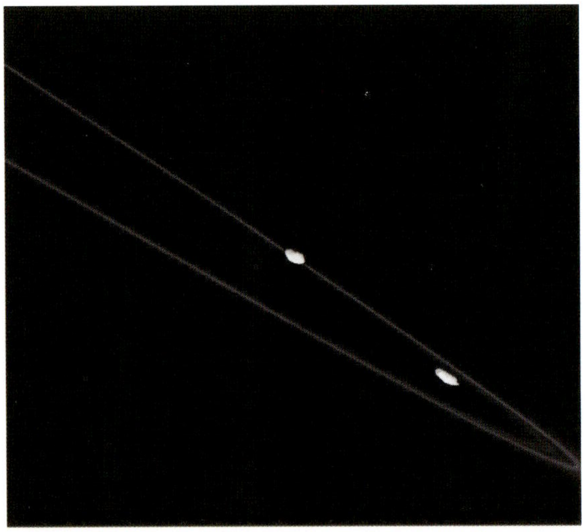

(a)

(b)

(a) This Cassini image shows the narrow, complex F Ring of Saturn, with its two small shepherd moons Pandora (left) and Prometheus (right). (b) In this closer view, the shepherd moon Pandora (84 kilometers across) is seen next to the F ring, in which the moon is perturbing the main (brightest) strand of ring particles as it passes. You can see the dark side of Pandora on this image because it is being illuminated by the light reflected from Saturn

Theoretical calculations suggest that the other narrow rings in the uranian and neptunian systems should also be controlled by shepherd moons, but none has been located. The calculated diameter for such shepherds (about 10 kilometers) was just at the limit of detectability for the Voyager cameras, so it is impossible to say whether they are present or not.

One of the outstanding problems with understanding the rings is determining their ages. Have the giant planets always had the ring systems we see today, or might these be a recent or transient addition to the solar system? In the case of the main rings of Saturn, their mass is about the same as that of the inner moon Mimas. Thus, they could have been formed by the break-up of a Mimas-sized moon, perhaps very early in solar system history, when there were many interplanetary projectiles left over from planet formation. It is harder to understand how such a catastrophic event could have taken place recently, when the solar system had become a more stable place.

Planetary System

planetary system, a star and all the celestial bodies bound to it by gravity, especially planets and their natural satellites . Until the last decade of the 20th cent., the only planetary system known was the solar system , which comprises the sun and the surrounding planets, natural satellites, asteroids, meteoroids, comets, and other celestial bodies. Speculation that other planetary systems exist dates back to antiquity, and through the years ever increasing numbers of astronomers searched for earthlike planets circling sunlike stars. The breakthrough came in 1992, when radio astronomers detected three planets orbiting a pulsar however, because pulsars are not normal stars, this was not considered a true planetary system. The first detection of an extrasolar planet, or exoplanet, around a normal star, 51 Pegasi, was made in 1995. This was quickly followed by the detection of a number of single exoplanets orbiting normal stars, and in 1999 the first discovery of multiple exoplanets orbiting a sunlike star, Upsilon Andromedae, was announced.

Because the solar system was the only planetary system known, all models of planetary systems were based on its characteristics several small planets close to the star, several large planets at greater distances, and circular planetary orbits. Most of the extrasolar planets discovered so far, however, are larger than earth, and many of those are much larger than Jupiter, the largest of the solar planets many of the rocky planets are much larger than earth, often up to 10 times more massive (one Kepler 10c, is 17 times as massive) many orbit their star at distances less than that of Mercury, the solar planet closest to the sun in one system found by the *Kepler* space telescope, five planets orbit a star more closely than Mercury does the sun and many have elliptical rather than circular orbits. Planets have also been found orbiting binary stars. All of this has caused planetary scientists to revisit accepted theories of planetary formation. Future theories will be measured against stars surrounded by a ring of gas and dust, such as Beta Pictoris, which are thought to be young adult stars with a planetary system forming around them.

An increasing number of planets with masses between one and seven times the earth's have been found in the 21st cent. In 2013 it was announced that astronomers reviewing the data from the *Kepler* space telescope had identified a planetary system (Kepler 37) that included a planet (Kepler 37c) somewhat smaller than the earth and another (Kepler 37b) that was smaller than Mercury. Kepler 37b is the smallest extrasolar planet discovered so far neither of the two planets was in the habitable zone. In 2014, Kepler scientists announced the discovery of a habitable-zone planet (Kepler 186f) with a radius estimated to be 10% larger than the earth's, that orbited a cool dwarf star with four other planets because of its size, Kepler 186f was believed to be a rocky planet with the potential to have liquid water. It is possible that some of the bodies that have been discovered are not planets in the solar-system sense but a new class of celestial bodies or even brown dwarfs .

Because stars are so distant and bright and an extrasolar planet, no matter how large, is relatively small and dim, it cannot be seen or photographed directly in visible light. Several techniques have been used to infer the presence of such planets. *Astrometry* is based on the slight gravitational disturbance, or wobble, that the planet causes in the motion of the star. *Photometry,* also called the *transit method,* is to measure the distinct dimming of light from the star as the planet's orbit brings it between the star and the earth. Using photometric techniques it also has been possible to photograph extrasolar planets in infrared light. *Doppler spectroscopy* is based on the fact that a planet periodically pulls its star closer to and farther from the earth as it orbits the star this motion has a measurable effect on the spectrum of light coming from the star. In *pulsar timing,* planets orbiting a pulsar can be detected by measuring the periodic variation in the pulse arrival time however, because the planets are orbiting a pulsar, a dead star, rather than a main-sequence star like the sun, this tends to be of less interest in the search for an earthlike extrasolar planet.

Heliocentrism

Historically, heliocentrism (that the sun was the centre of the universe) was opposed to geocentrism, which placed the Earth at the centre of the universe. The notion of a heliocentric solar system, with the Sun at the centre, is possibly first suggested in the Vedic literature of ancient India, which often refer to the Sun as the "centre of spheres". Some interpret Aryabhatta's writings in Āryabhaṭīya as implicitly heliocentric. The idea was first proposed in western philosophy and Greek astronomy as early as the 3rd century BC by Aristarchus of Samos, but had received no support from most other ancient astronomers.

Speculation on Extrasolar Planetary Systems

In the 16th century the Italian philosopher Giordano Bruno, an early supporter of the Copernican theory that the Earth and other planets orbit the Sun, put forward the view that the fixed stars are similar to the Sun and are likewise accompanied by planets. He was burned at the stake for his ideas by the Roman Inquisition.

In the 18th century the same possibility was mentioned by Isaac Newton in the "General Scholium" that concludes his Principia. Making a comparison to the Sun's planets, he wrote "And if the fixed stars are the centres of similar systems, they will all be constructed according to a similar design and subject to the dominion of One. His theories gained traction through the 19th and 20th centuries despite a lack of supporting evidence. Long before their confirmation by astronomers, conjecture on the nature of planetary systems had been a focus of Search for extraterrestrial intelligence and has been a prevalent theme in fiction, particularly science fiction.

Detection of Exoplanets

The first confirmed detection of a planetary system was in 1992, with the discovery of several terrestrial-mass planets orbiting the pulsar PSR B1257+12. The first confirmed detection of planetary system of a main-sequence star was made in 1995, when a giant planet, 51 Pegasi b, was found in a four-day orbit around the nearby G-type star 51 Pegasi. The frequency of detections has increased since then, particularly through advancements in methods of detecting extrasolar planets and dedicated planet finding programs such as the Kepler mission.

An artist's concept of a protoplanetary disk. Image from Smithsonian

Origin and Evolution

Early theories of planetary systems were based on knowledge of the formation and evolution of the Solar System. Planetary systems are generally believed to form as part of the same process which results in star formation. Some early theories involved another star passing extremely close to the star, drawing material out from it which then coalesced to form the planets. However, the probability of such a near collision is now known to be far too low to make this a viable model. Accepted theories today argue that a protoplanetary disk forms by gravitational collapse of a molecular cloud and then evolves into a planetary system by collisions and gravitational capture.

Some planetary systems may form differently, however. Planets orbiting pulsars stars which emit periodic bursts of electromagnetic radiation have been discovered by the slight variations they cause in the timing of these bursts. Pulsars are formed in violent supernova explosions, and a normal planetary system could not possibly survive such a blast planets would either evaporate, be pushed off of their orbits by the masses of gas from the exploding star, or the sudden loss of most of the mass of the central star would see them escape the gravitational hold of the star. One theory is that existing stellar companions were almost entirely evaporated by the supernova blast, leaving behind planet-sized bodies. Alternatively, planets may somehow form in the accretion disk surrounding pulsars.

The formation of planetary systems has also been found to be linked to stellar classification. Stars are composed mainly of the light elements hydrogen and helium. They also contain a small proportion of heavier elements such as iron, and this fraction is referred to as a star's metallicity. Stars of higher metallicity are much more likely to form and retain planetary systems, and the planets within them tend to be more massive than those of lower-metallicity stars. It has also been shown that stars with planets are more likely to be deficient in lithium. Recent observations by the Spitzer Space Telescope indicate that planetary formation does not occur around other stars in the vicinity of an O class star due to the photoevaporation effect.

Models of the Solar System point to material being ejected during the formation and early evolution as the system stabilised. Similar interactions are thought to be common to many planetary systems. The orbits of large bodies are being established; they may migrate and cause planetary body collisions. Others may be sufficiently perturbed to be ejected completely from the system. These ejected bodies are known as rogue planets and may retain their natural satellites. Systems such as these in interstellar space, known as planetars, may orbit the galaxy directly and although difficult to detect are now thought to be extremely common.

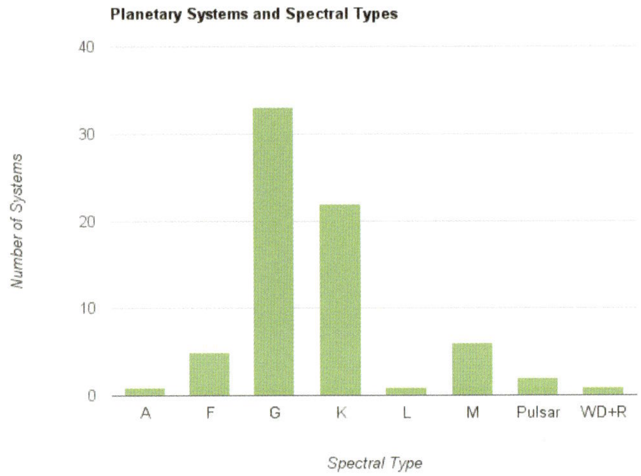

Planetary systems

Types and Attributes

With the increasing discovery of planetary systems, astronomers have begun to classify them by type. In particular, as certain types of stars are now known for producing specific types of planetary systems, they are classified by the spectral type of the host star. Main-sequence stars like the Sun (spectral type G or K), for example, have represented the majority of planetary system discoveries. Additionally, they are often classified by the size and types of planets and their orbital configuration. Hot Jupiter systems with a gas giant very close to the star are one of the most common found so far and Hot Neptune type systems have also been found. Theories, such as scattering, have been proposed for the formation of large planets close to their parent stars. Dusty disks with a wide ring of dust and comets are another common type of system. Protoplanetary disks, still in the process of forming have also been discovered. At present, few systems have been found to analogues of our own with terrestrial planets close to the parent star. More commonly, systems consisting of multiple Super-Earths have been detected.

Multiplanet Systems

140 systems (including the Solar System) are now known to consist of two or more planets. In terms of total confirmed planets, the Solar System with 8 remains the largest planetary system and HD 10180 with 7, the second largest. 55 Cancri was for some time the largest known system. Some systems, such as HD 10180 and Gliese 581 contain multiple unconfirmed planets pushing the total expected number of planets to be discovered in their systems higher. If HD 10180's two candidate planets are confirmed, it would be the largest by number of known planets with a total of 9.

Orbital Properties

Unlike the Solar System, which has coplanar and circular orbits, many of the known planetary systems display much higher orbital eccentricity and some consist of planets in dramatic elliptical orbits while some consist of planets on inconsistent orbital inclination. An example of such a system is 16 Cygni. Although no planets in the Solar System are completely tidally locked, planets with regular orbits in close to their primary stars are thought to typically experience tidal locking. aaSome systems also appear to display orbital resonances. Gliese 876, for example has a 3-body Laplace resonance. Kepler-9 has two planets in mean motion resonance, the first transiting planets discovered with such a resonance.

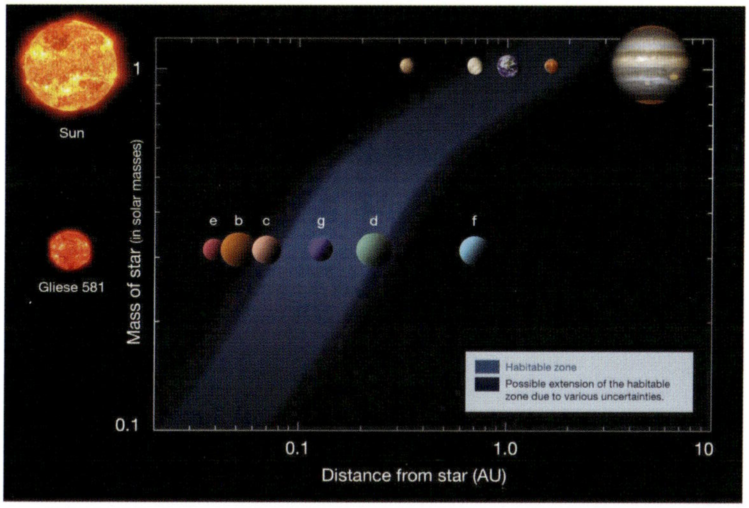

The habitable zone of Gliese 581 compared with our Solar System's habitable zone

Habitable Zones

The habitable zone is the region in the planetary system where an Earth-like planet (or moon) could maintain liquid water on its surface and potentially therefore, life. The distance of this zone from the parent star can vary greatly from system to system. The lack of surface water on Mars, Venus, the Moon, and Ceres, which according to some estimates lie within the habitable zone of the Solar System, does not exclude the possibility of extraterrestrial surface life in the Solar System but makes it highly unlikely. However the abundance of Life on Earth makes the habitable zones of exoplanetary systems of particular interest to exobiologists and the Search for extraterrestrial intelligence. Gliese 581 was the first exoplanetary system found with Super-Earths in the habitable zone. HD 37124 has multiple known planets in the zone. Systems with gas giants have more commonly discovered in the habitable zone, though they could possibly harbour Earth-like natural satellites.

Main Sequence Systems

The Solar System was the first planetary system known around a main-sequence star. As such, main-sequence systems are of particular interest to science due to the prospect of finding true analogs to the Solar System, along with true Earth analogs and therefore possibly Earth-like complex or even intelligent life.

The first main-sequence exoplanetary system found was Upsilon Andromedae in April 1999. Since then, several further systems have been discovered. Many have been found to bear similarities to our own, for example the first terrestrial planets of comparable size to Earth (in the Kepler-20 system) and planets close to Earth sized but much larger (Super-Earths) have even been located in habitable zones in the HD 85512 and Kepler 22 systems. Discoveries to date indicate that our Solar System's arrangement of all the terrestrial planets arranged closest to the sun and the giant planets further out is unlikely to be the general rule. This challenges well established theories of their formation and evolution suggesting that early migration of giant planets could causes wide variations in the arrangement of main-sequence systems.

Three of the nearest main-sequence stars to the Sun are now thought to host planetary systems. This includes the nearest main-sequence star to Earth at 4.37ly, Alpha Centauri A which was found in 2012 to have an Earth-sized planet in close orbit around it. Unconfirmed planetary systems include the Epsilon Eridani system at 10.5ly from Earth and the Tau Ceti system at 11.905ly, both believed to have multiple detected planets.

This artist's concept illustrates a young, red dwarf star surrounded by three planets

Red Dwarf Systems

Red dwarf systems orbit red dwarfs, stars which may represent 70% to more than 90% of the stars in the galaxy. The Gliese 876 was the first such system discovered in 1998. Most planets of these systems are thought to be tidally locked. This and high stellar variation are postulated to be major impediments to planetary habitability. Gliese 581 is another well known red dwarf system which is speculated to have multiple planets within the habitable zone.

Brown Dwarf Systems

Brown dwarf planetary system orbit brown dwarfs. 2M1207, the first planetary system to be imaged, was also the first such system discovered. Some companions of brown dwarfs, such as those of 2M1207 and GQ Lupi, are thought to have formed by cloud collapse rather than accretion, and hence these are sometimes considered sub-brown dwarfs rather than planets. Disks around brown dwarfs have been found to have many of the same features as disks around stars; therefore, it is expected that there will be accretion-formed planets around brown dwarfs. Given the small mass of brown dwarf disks, most planets will be terrestrial planets rather than gas giants. It is now believed that many brown dwarfs may have planets and also, possibly, a shortly lived habitable zone. Most planets of these systems are thought to be tidally locked.

Sub-brown Dwarf Systems

Rogue systems are systems that are not gravitationally bound to any star and that therefore orbit the galaxy directly. Some astronomers have estimated that there may be twice as many Jupiter-sized rogue planets as there are stars. Isolated planetary-mass objects which were not ejected from a star's orbit, may have formed in a similar way to stars. The IAU has proposed that those objects be called sub-brown dwarfs. It is theorised that objects ejected from a planetary system could retain a system of satellites. Cha 110913-773444 is an example of a rogue system with a protoplanetary disk speculated to contain as many as 4 satellites. It is unknown whether it is ejected planet, or formed on its own.

Pulsar Systems

Pulsar systems orbit pulsars, or rapidly rotating neutron stars. PSR B1257+12 was not only the first pulsar exoplanetary system but also the first planetary system discovered. PSR B1620-26 is also confirmed, however many other pulsar systems remain unconfirmed. The formation of pulsar systems is still the subject of research. It is thought that planets form from the debris of companion stars.

An artist's impression of the binary star system HD 98800 B, which is surrounded by a disc that may be in the process of forming planets. HD 98800 B is itself a member of a quadruple star system

Multiple Star Systems

Multiple star systems, including binary stars are abundant in the Milky Way Galaxy and several are now known to possess planetary systems. The discovery of such systems however creates new challenges on theories on planetary system formation. Prior to the groundbreaking 1993 discovery of the first known planet PSR B1620-26 b in the binary system PSR B1620-26, the possibility of the existence of such systems was long theorised and debated. Chinese-American astrophysicist Su-Shu Huang had dampened expectations of planets forming or remaining in such systems in his 1959 paper "Life-Supporting Regions in the Vicinity of Binary Systems" claiming that due to orbital instability planets and in particular habitable planets would likely be rare. The film Star Wars, however captured the imagination of popular culture with its "twin sun" scene on Tatooine.

PSR B1620-26 was not only the first binary system (and multistar system) found with a planet but the first found in which the planet orbited both stars (i.e. in circumbinary orbit, a circumbinary planet).

16 Cygni, discovered in 1996 was the first confirmed triple star planetary system found however the planet orbits just one of the stars – 16 Cygni B. In 2011, the Kepler-16 system was found to have a circumbinary planet within its habitable zone, leading to the Smithsonian Centre informally referring to it as "Tatooine" in reference to Star Wars.

This type of planetary system is of particular interest due to Alpha Centauri A being the closest main-sequence star to Earth but also a binary (Alpha Centauri AB) contained within a triple star system. The notion of planets in the Centauri system has been a popular theme in fiction. In 2012 Alpha Centauri B was found to have an Earth-sized planet in orbit around it, increasing expectations of more terrestrial planet discoveries in this and similar systems. PH1, confirmed in 2012, is the first quadruple star planetary system confirmed. NN Serpentis and HW Virginis are notable multi-planet multi-star systems.

Transiting Systems

Transiting planetary systems are those with multiple planets in orbital planes along the line of sight between their star and Earth. The transit method is one of the most successful Earth-centric observation methods. The Kepler mission was designed to discover systems of this type and has found numerous candidate systems. Kepler-11 is a prominent example of a transiting system in which multiple simultaneous planetary transits have been observed. Transiting systems can be studied to not only discover planets, but through Astronomical spectroscopy and Transit timing effects accurately analyse the properties of the systems including the detection of extraterrestrial atmospheres, dust or asteroid belts and possibly natural satellites.

System Architectures

The Solar System consists of an inner region of small rocky planets and outer region of large gas giants. However, other planetary systems can have quite different architectures. Studies suggest that architectures of planetary systems are dependent on the conditions of their initial formation. Many systems with a hot Jupiter gas giant very close to the star have been found. Theories, such as planetary migration or scattering, have been proposed for the formation of large planets close to their parent stars. At present, few systems have been found to be analogous to the Solar System with terrestrial planets close to the parent star. More commonly, systems consisting of multiple Super-Earths have been detected.

Components

Planets

Some studies suggest that there is at least one planet on average per star. This would suggest that, like the Solar System, most stars have planets (or exoplanets). However, the proportion of stars is uncertain because not all planets can yet be detected. The radial-velocity method and the transit method (which between them are responsible for the vast majority of detections) are most sensitive to large planets in small orbits. Thus, many known exoplanets are "hot Jupiters": planets of Jovian mass or larger in very small orbits with periods of only a few days. A 2005 survey of radial-velocity-detected planets found that about 1.2% of Sun-like stars have a hot Jupiter, where "Sun-like star" refers to any main-sequence star of spectral classes late-F, G, or early-K without a close stellar companion. This 1.2% is more than double the frequency of hot jupiters detected by the Kepler spacecraft, which may be because the Kepler field of view covers a different region of the Milky Way where the metallicity of stars is different. It is further estimated that 3% to 4.5% of Sun-like stars possess a giant planet with an orbital period of 100 days or less, where "giant planet" means a planet of at least 30 Earth masses.

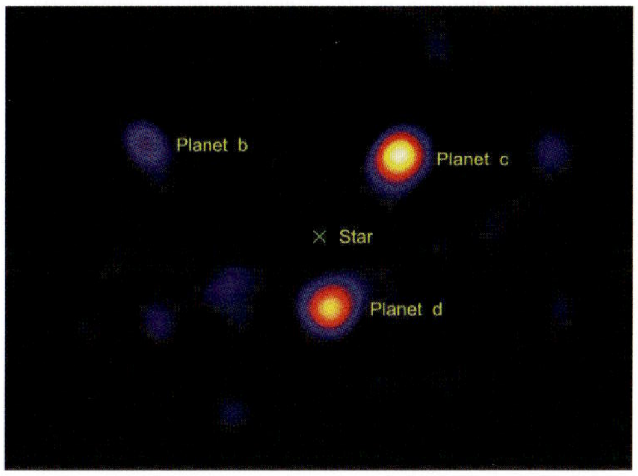

The three known planets of the star HR8799, as imaged by the Hale Telescope. The light from the central star was blanked out by a vector vortex coronagraph

It is known that small planets (of roughly Earth-like mass or somewhat larger) are more common than giant planets. It also appears that there are more planets in large orbits than in small orbits. Based on this, it is estimated that perhaps 20% of Sun-like stars have at least one giant planet, whereas at least 40% may have planets of lower mass. A 2012 study of gravitational microlensing data collected between 2002 and 2007 concludes the proportion of stars with planets is much higher and estimates an average of 1.6 planets orbiting between 0.5–10 AU per star in the Milky Way. The authors of this study conclude that "stars are orbited by planets as a rule, rather than the exception".

Whatever the proportion of stars with planets, the total number of exoplanets must be very large. Because the Milky Way has at least 200 billion stars, it must also contain tens or hundreds of billions of planets.

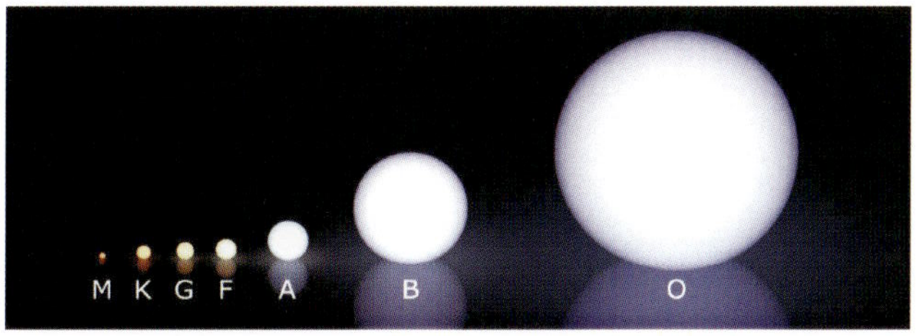

The Morgan-Keenan spectral classification

Most known exoplanets orbit stars roughly similar to the Sun, that is, main-sequence stars of spectral categories F, G, or K. One reason is that planet-search programs have tended to concentrate on such stars. In addition, statistical analyses indicate that lower-mass stars (red dwarfs, of spectral category M) are less likely to have planets massive enough to be detected by the radial-velocity method. Nevertheless, several tens of planets around red dwarfs have been discovered by the Kepler spacecraft by the transit method, which can detect smaller planets.

Stars of spectral categories A and B typically rotate very quickly, which makes it very difficult to measure the small Doppler shifts induced by orbiting planets because the spectral lines are very broad. However, this type of massive star eventually evolves into a cooler red giant that rotates more slowly and thus can be measured using the radial-velocity method. A few tens of planets have been found around red giants.

Observations using the Spitzer Space Telescope indicate that extremely massive stars of spectral category O, which are much hotter than the Sun, produce a photo-evaporation effect that inhibits planetary formation. When the O-type star goes supernova any planets that had formed would become free-floating due to the loss of stellar mass unless the natal kick of the resulting remnant pushes it in the same direction as an escaping planet. Fallback disks of matter that failed to escape orbit during a supernova may form planets around neutron stars and black holes.

Doppler surveys around a wide variety of stars indicate about 1 in 6 stars having twice the mass of the Sun are orbited by one or more Jupiter-sized planets, vs. 1 in 16 for Sun-like stars and only 1 in 50 for red dwarfs. On the other hand, microlensing surveys indicate that long-period Neptune-mass planets are found around 1 in 3 red dwarfs. Kepler Space Telescope observations of planets with up to one year periods show that occurrence rates of Earth- to Neptune-sized planets (1 to 4 Earth radii) around M, K, G, and F stars are successively higher towards cooler, less massive stars.

At the low-mass end of star-formation are sub-stellar objects that don't fuse hydrogen: the brown dwarfs and sub-brown dwarfs, of spectral classification L,T and Y. Planets and protoplanetary disks have been discovered around brown dwarfs, and disks have been found around sub-brown dwarfs (e.g. OTS 44).

Ordinary stars are composed mainly of the light elements hydrogen and helium. They also contain a small proportion of heavier elements, and this fraction is referred to as a star's metallicity (even if the elements are not metals in the traditional sense), denoted [m/H] and expressed on a logarithmic scale where zero is the Sun's metallicity.

A 2012 study of the Kepler spacecraft data found that smaller planets, with radii smaller than Neptune's were found around stars with metallicities in the range $-0.6 < [m/H] < +0.5$ (about four times less than that of the Sun to three times more), whereas larger planets were found mostly around stars with metallicities at the higher end of this range (at solar metallicity and above). In this study small planets occurred about three times as frequently as large planets around stars of metallicity greater than that of the Sun, but they occurred around six times as frequently for stars of metallicity less than that of the Sun. The lack of gas giants around low-metallicity stars could be because the metallicity of protoplanetary disks affects how quickly planetary cores can form and whether they accrete a gaseous envelope before the gas dissipates. However, Kepler can only observe planets very close to their star and the detected gas giants probably migrated from further out, so a decreased efficiency of migration in low-metallicity disks could also partly explain these findings.

A 2014 study found that not only giant planets, but planets of all sizes, have an increased occurrence rate around metal-rich stars compared to metal-poor stars, although the larger the planet, the greater this increase as the metallicity increases. The study divided planets into three groups based on radius: gas giants, gas dwarfs, and terrestrial planets with the dividing lines at 1.7 and 3.9 Earth radii. For these three groups, the planet occurrence rates are 9.30, 2.03, and 1.72 times higher for metal-rich stars than for metal-poor stars, respectively. There is a bias against detecting smaller planets because metal-rich stars tend to be larger, making it more difficult to detect smaller planets, which means that these increases in occurrence rates are lower limits.

It has also been shown that stars with planets are more likely to be deficient in lithium.

Most stars form in open clusters, but very few planets have been found in open clusters and this led to the hypothesis that the open-cluster environment hinders planet formation. However, a 2011 study concluded that there have been an insufficient number of surveys of clusters to make such a hypothesis. The lack of surveys was because there are relatively few suitable open clusters in the Milky Way. Recent discoveries of both giant planets and low-mass planets in open clusters are consistent with there being similar planet occurrence rates in open clusters as around field stars. The open cluster NGC 6811 contains two known planetary systems Kepler-66 and Kepler-67.

Circumstellar Disks and Dust Structures

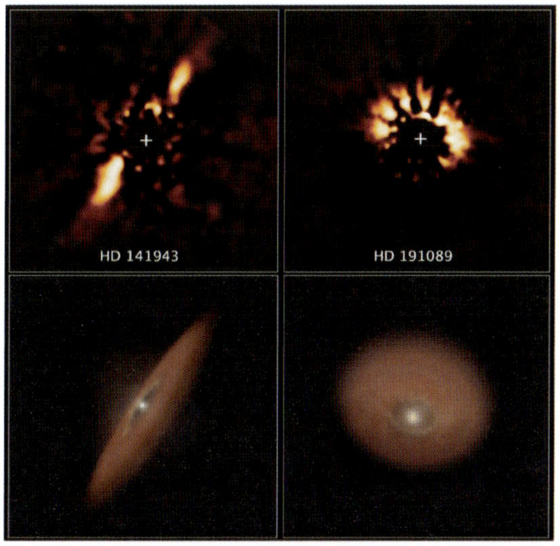

Debris disks detected in HST archival images of young stars, *HD 141943* and *HD 191089*, using improved imaging processes

After planets, circumstellar disks are one of the most commonly observed properties of planetary systems, particularly of young stars. The Solar System possesses at least four major circumstellar disks (the asteroid belt, Kuiper belt, scattered disc, and Oort cloud) and clearly observable disks have been detected around nearby solar analogs including Epsilon Eridani and Tau Ceti. Based on observations of numerous similar disks, they are assumed to be quite common attributes of stars on the main sequence.

Interplanetary dust clouds have been studied in the Solar System and analogs are believed to be present in other planetary systems. Exozodiacal dust, an exoplanetary analog of zodiacal dust, the 1–100 micrometre-sized grains of amorphous carbon and silicate dust that fill the plane of the Solar System has been detected around the 51 Ophiuchi, Fomalhaut, Tau Ceti, and Vega systems.

Comets

There are 5,253 known Solar System comets and they are thought to be common components of planetary systems. The first exocomets were detected in 1987 around Beta Pictoris, a very young A-type main-sequence star. There are now a total of 11 stars around which the presence of exocomets have been observed or suspected. All discovered exocometary systems, are around very young A-type stars.

Other Components

Computer modeling of an impact in 2013 detected around the star NGC 2547 by the Spitzer Space Telescope and confirmed by ground observations suggests the involvement of large asteroids or protoplanets similar to the events believed to have led to the formation of terrestrial planets like the Earth.

Based on observations of the Solar System's large collection of natural satellites, they are believed common components of planetary systems; however, exomoons have so far eluded confirmation. The star 1SWASP J140747.93-394542.6, in the constellation Centaurus, is a strong candidate for a natural satellite. Indications suggest that the confirmed extrasolar planet WASP-12b also has at least one satellite.

Orbital Configurations

Unlike the Solar System, which has orbits that are nearly circular, many of the known planetary systems display much higher orbital eccentricity. An example of such a system is 16 Cygni.

Mutual Inclination

The mutual inclination between two planets is the angle between their orbital planes. Many compact systems with multiple close-in planets interior to the equivalent orbit of Venus are expected to have very low mutual inclinations, so the system (at least the close-in part) would be even flatter than the solar system. Captured planets could be captured into any arbitrary angle to the rest of the system. The only system where mutual inclinations have actually been measured is the Upsilon Andromedae system: the planets, c and d, have a mutual inclination of about 30 degrees.

Orbital Dynamics

Planetary systems can be categorized according to their orbital dynamics as resonant, non-resonant-interacting, hierarchical, or some combination of these. In resonant systems the orbital periods of the planets are in integer ratios. The Kepler-223 system contains four planets in an 8:6:4:3 orbital resonance. Giant planets are found in mean-motion resonances more often than smaller planets. In interacting systems the planets orbits are close enough together that they perturb the orbital parameters. The Solar System could be described as weakly interacting. In strongly interacting systems Kepler's laws do not hold. In hierarchical systems the planets are arranged so that the system can be gravitationally considered as a nested system of two-bodies, e.g. in a star with a close-in hot jupiter with another gas giant much further out, the star and hot jupiter form a pair that appears as a single object to another planet that is far enough out.

Other, as yet unobserved, orbital possibilities include: double planets; various co-orbital planets such as quasi-satellites, trojans and exchange orbits; and interlocking orbits maintained by precessing orbital planes.

Number of Planets, Relative Parameters and Spacings

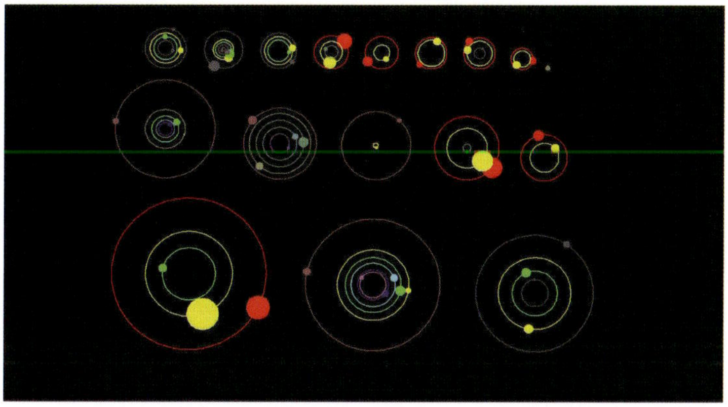

The spacings between orbits vary widely amongst the different systems discovered by the Kepler spacecraft

- On The Relative Sizes of Planets Within Kepler Multiple Candidate Systems, David R. Ciardi
- The Kepler Dichotomy among the M Dwarfs: Half of Systems Contain Five or More Coplanar Planets, Sarah Ballard, John Asher Johnson
- Exoplanet Predictions Based on the Generalised Titius-Bode Relation, Timothy Bovaird, Charles H. Lineweaver
- The Solar System and the Exoplanet Orbital Eccentricity - Multiplicity Relation, Mary Anne Limbach, Edwin L. Turner
- The period ratio distribution of Kepler's candidate multiplanet systems, Jason H. Steffen, Jason A. Hwang
- Are Planetary Systems Filled to Capacity? A Study Based on Kepler Results, Julia Fang, Jean-Luc Margot

Planet Capture

Free-floating planets in open clusters have similar velocities to the stars and so can be recaptured. They are typically captured into wide orbits between 100 and 10^5 AU. The capture efficiency decreases with increasing cluster size, and for a given cluster size it increases with the host/primary mass. It is almost independent of the planetary mass. Single and multiple planets could be captured into arbitrary unaligned orbits, non-coplanar with each other or with the stellar host spin, or pre-existing planetary system. Some planet–host metallicity correlation may still exist due to the common origin of the stars from the same cluster. Planets would be unlikely to be captured around neutron stars because these are likely to be ejected from the cluster by a pulsar kick when they form. Planets could even be captured around other planets to form free-floating planet binaries. After the cluster has dispersed some of the captured planets with orbits larger than 10^6 AU would be slowly disrupted by the galactic tide and likely become free-floating again through encounters with other field stars or giant molecular clouds.

Small Solar System Body

Small body, also called small solar system body (SSSB), any natural solar system object other than the Sun and the major planets and dwarf planets and their satellites (moons). The small bodies populate the solar system in vast numbers and include the mostly rocky asteroids, or minor planets, the predominantly icy comets, and the fragments of such bodies commonly called meteoroids over a continuum of sizes down to microscopic grains known as interplanetary dust particles or micrometeoroids. The term small body is sometimes restricted to objects that can be observed telescopically while still outside Earth's atmosphere. In practice, that results in a lower limit of a few metres on the diameters of small bodies.

Composite image of the nucleus of Comet Halley

The use of meteoroid has expanded to describe any small chunk of matter in interplanetary space, especially one less than a few tens of metres in size. The term formerly was limited to small bodies that were near enough to Earth to eventually enter its atmosphere. In that narrower meaning, meteoroid is usually linked with the terms meteor, meteorite, or both. As a meteoroid travels through the atmosphere, friction heats it up, creating a glowing trail of hot ionized gases (plasma) called a meteor. If the object survives both its passage through the atmosphere and its subsequent impact on the ground, it is called a meteorite. Over 99 percent of meteorites are fragments of asteroids. A small group is known to be of lunar origin, and a second group is generally acknowledged to have come from Mars. There is also reason to believe that some are fragments of the rocky remnants of comets, although that remains to be firmly established.

The Oort Cloud is a spherical shell of millions of icy bodies which surrounds the solar system at vast distances and is thought to be the birth place of long-period comets.

The Kuiper Belt is a region extending from Neptune's orbit out to the far and distant reaches of the solar system and possibly holds the best available record of the original interstellar materials that formed the solar nebula. This region beyond Neptune is also the most probable birthplace of the short-period comets.

Comets are pristine remnants from the formation of the solar system that are comprised of minerals, rock and mostly ice, much like a dirty snowball. They travel around the sun in elliptical orbits and can be inclined to the plane of the solar system at any angle. Comets can sprout tails extending many tens of millions of miles, during their closest approach to the sun. Short period comets are thought to come from the Kuiper Belt on the outskirts of Neptune's orbit and further, and longer period comets are thought to come from the Oort cloud, a vast spherical shell that surrounds the solar system at a huge distance. Recent spacecraft encounters with comets seem to raise more questions then they answer and some finds are quite unexpected. NASA targets some of these bodies with spacecraft loaded with instrumentation that help tease out the secrets lurking in these icy bodies.

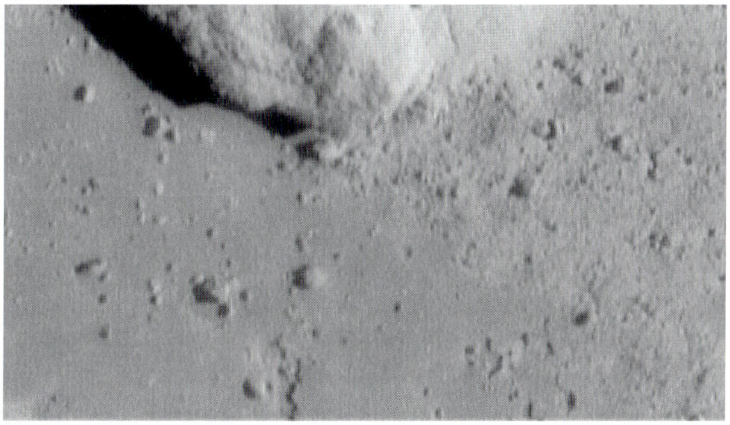

Pluto resides in the Kuiper Belt. With an orbit inclined to the plane of the solar system, Pluto most likely evolved away from the sun's flattened disk where the larger bodies (or planets) formed. Pluto's orbit crosses inside that of Neptune's rendering Pluto also a member of the Trans-Neptunian Object (TNO) class. The Pluto system is very exotic, having three moons including Charon discovered in 1978, and Nix and Hydra discovered in 2005.

Asteroids are rocky remnants from the formation of the solar system. They are not spherical and have differing compositions and histories. Most, although not all asteroids, reside in a region between Mars and Jupiter where numerous other small rocky worlds orbit the sun. Some asteroids belong to groups that came from larger parent bodies which were shattered in past collisions with other asteroids. Some are in orbits that cross paths with that of Earth's or other planets. Asteroids that cross Earth's orbit are called Potentially Hazardous Asteroids (PHA) and the more we observe the heavens, the more of them we find, some of which are seen for the first time just after passing close to Earth.

Populations

The largest known small bodies, in the conventional sense, are several icy Kuiper belt objects found orbiting the Sun beyond the orbit of Neptune. Ceres which is the largest main-belt asteroid and is now considered a dwarf planet is roughly 950 km (590 miles) in diameter. The distant

Kuiper belt object Eris, roughly the same size as Pluto, which has a diameter of about 2,350 km (1,460 miles), is also designated a dwarf planet rather than a small body, as are Haumea and Makemake, both of which have diameters of about 1,450 km (900 miles). These four dwarf planets are known as plutoids. At least four other Kuiper belt objects are also believed to be larger than Ceres. In the 1990s astronomers recognized that Pluto, which has a diameter of about 2,300 km (1,400 miles), is a giant icy leftover from the time when the planets formed. Pluto had been classified as a planet from its discovery in 1930 but was reclassified as a dwarf planet in 2006. Primarily for historical reasons, the moons of the planets are not considered small bodies. Nevertheless, many moons are asteroid-size or smaller, and some are probably captured asteroids or comet nuclei.

Although small bodies are found throughout the solar system, most of those known are concentrated in several regions. Those regions in which they travel in fairly stable orbits include (1) the main asteroid belt, between the orbits of Mars and Jupiter, (2) the gravitationally stable points (Lagrangian points) along the orbital paths of Jupiter, Mars, Uranus, Neptune, and Earth where objects called Trojan asteroids are clustered, (3) the Kuiper belt, a disk-shaped zone of icy bodies whose members are also called trans-Neptunian objects because they all have orbits beyond that of Neptune, and (4) the spherical Oort cloud of icy bodies inferred to be orbiting the Sun at distances typically more than 1,000 times that of Neptune or Pluto. In none of the above categories is Sedna, an object discovered in 2003 in the outer reaches of the solar system that is perhaps as large as 40 percent the size of Pluto. Sedna has a highly elliptical orbit in which its closest approach to the Sun is more than twice that of Neptune i.e., billions of kilometres beyond the recognized extent of the Kuiper belt; it may be a member of a population of objects lying between the Kuiper belt and the Oort cloud.

Other small bodies travel in unstable orbits that cross the paths of one or more of the planets. Those include (1) most observed comets, (2) near-Earth asteroids, most with orbits that cross either Earth's orbit or Mars's orbit and some with orbits that lie mostly or entirely inside Earth's orbit and cross the orbits of Venus or of both Venus and Mercury while closely approaching Earth's, (3) Centaur objects, icy bodies that are thought to have been gravitationally perturbed out of the Kuiper belt and now travel mainly between the orbits of Jupiter and Neptune, and (4) individual idiosyncratic planet-crossing objects such as the asteroid Hidalgo, which travels between the inner edge of the asteroid belt and a point just beyond the orbit of Saturn. All objects in planet-crossing orbits eventually collide with the Sun or a planet or are permanently ejected from the solar system, although some survive for long periods (up to hundreds of millions of years) because of stabilizing resonances.

Essentially all the small bodies are thought to be remnant material from the planet-building process that took place during the formation of the solar system from the solar nebula. Near the hot nascent Sun, closer than Jupiter's present orbit, only the less-volatile high-temperature-freezing-point substances such as metals and silicates could condense from the nebula. Beginning as tiny grains, that material gradually accreted under mutual gravitational influence to become large rocky bodies up to hundreds of kilometres in diameter. Most of those large objects, called planetesimals, ultimately combined to form the dense rocky planets of the inner solar system (Mercury through Mars). In the outer solar system, near Jupiter's present orbit and beyond, temperatures were cool enough to allow abundant volatile substances such as water and carbon dioxide to condense to their ices. Although rocky planetesimals likewise accreted there, the significant additional mass contributed by the ices allowed the growth of the low-density gas-rich outer planets (Jupiter through Neptune).

Nevertheless, in the inner and the outer solar system, there remained rocky bodies and predominantly icy bodies, respectively, that were never incorporated into planets or captured as planetary satellites. It is now believed that many of these bodies were subsequently mixed together and relocated from their formation locations by the migration of the giant planets. Today, in the inner solar system, these remnants are represented chiefly by the asteroids and their fragments; in the outer solar system, by the objects in and derived from the Kuiper belt and the Oort cloud. Comets originate, and most are still located, in the Kuiper belt and the Oort cloud. Except for the very largest Kuiper belt objects (diameters greater than about 100 km [60 miles]), comets in the Kuiper belt and beyond are unobservable because of their great distances from Earth. It is not until they enter the realm of the planets that they develop comas and tails.

Nice Model

"Nice model" describes dynamical evolution of the outer Solar System since the time when the gas was removed from the protoplanetary disk. In this model, the giant planets underwent a dynamical instability which played a major role in shaping the present-day Solar System. This instability could have happened several 100 million years after the planets formed. The Nice model can explain several observations in the Solar System, including the orbits of the giant planets, the existence and the orbital structure of several small body populations (e.g., Jupiter's Trojans), and the late heavy bombardment.

The best understanding of how the planets have moved in throughout our system's evolution arose from the Nice Model, proposed by an international collaboration of scientists in 2005. This model suggests that at the inner edge of the icy disk, some 35 AU from the Sun, the outermost planet began interacting with icy planetesimals, influencing the second sort of migration to occur: gravitational scattering. Comets were slingshotted from one planet to the next, which gradually caused Uranus, Neptune, Saturn and the belt to migrate outwards. Jupiter's powerful gravity flung the icy objects that reached it into highly elliptical orbits or out of the Solar System entirely, which in order to conserve angular momentum, further propelled its journey inwards.

An extension to this theory is the 'Grand Tack model', which is named after the unusual course of Jupiter's migration towards the Sun before stopping and migrating outwards again, like a sailboat tacking about a buoy. At the distance that Mars would later coalesce, material had been swept away due to Jupiter's presence. This resulted in the stunted growth of Mars and a material-rich region from which the Earth and Venus formed, explaining their respective sizes. The gas giant's travels also prevented the rocky material in the asteroid belt from accreting into larger bodies due to its strong gravitational influence. Although Jupiter swapped positions with the asteroid belt twice the movements were so slow that collisions were minimal, resulting in more of a gentle displacement.

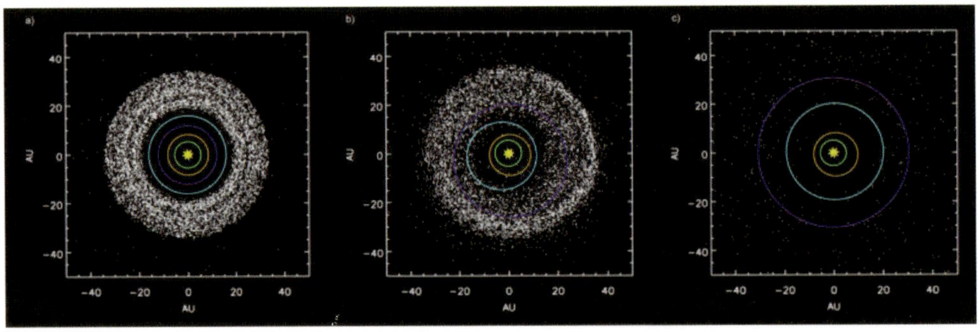

The Nice Model

But why did Jupiter's migration to the Sun's fiery depths cease? For that it has Saturn to thank. As the two planets moved further away from each other, it was believed they became temporarily locked in a 2:1 orbital resonance. That meant that for every orbit of the Sun Saturn made, Jupiter made two. The Nice Model showed that the planetary coupling increased their orbital eccentricities and rapidly destabilised the entire system. Jupiter forced Saturn outwards, pushing Neptune and Uranus into extremely elliptical orbits where they gravitationally scattered the dense icy disk far into the inner and outer Solar System. This disruption in turn scattered almost the entire primordial disk. Some models also show Neptune to have been propelled past Uranus to become the farthest planet from the Sun as we now know it. Over time the orbits of the outermost planets settled back into the near circular paths we observe today.

The Nice Model explains the present day absence of a dense trans-Neptunian population and the positions of the Kuiper belt and Oort cloud. It also accounts for the mixture of icy and rocky objects in the asteroid belt, like water-rich dwarf planet, Ceres, which likely originated from the icy belt. The rapid scattering of icy objects, around 4 billion years ago, dates with the onset of the late heavy bombardment period, which is predominantly recorded from the Moon's well-preserved surface.

References

- Newton, Isaac; Cohen, I. Bernard; Whitman, Anne (1999) [First published 1713]. The Principia: A New Translation and Guide. University of California Press. p. 940. ISBN 0-520-20217-1
- Small-bodies-solar-system, focus-areas, solar-system: science.nasa.gov, Retrieved 23 May 2018
- Stuart J. Weidenschilling & Francesco Marzari (1996). "Gravitational scattering as a possible origin for giant planets at small stellar distances". Nature. 384 (6610): 619–621. Bibcode:1996Natur.384..619W. doi:10.1038/384619a0. PMID 8967949
- Small-body, science: britannica.com, Retrieved 18 July 2018
- L. Vu (3 October 2006). "Planets Prefer Safe Neighborhoods". Spitzer Science Center. Archived from the original on July 13, 2007. Retrieved 2007-09-01
- Stark, C..; Kuchner, M. (2008). "The Detectability of Exo-Earths and Super-Earths Via Resonant Signatures in Exozodiacal Clouds". The Astrophysical Journal. 686 (1): 637–648. arXiv:0810.2702. Bibcode:2008ApJ...686..637S. doi:10.1086/591442
- How-are-planetary-rings-formed: blogstronomy.blogspot.com, Retrieved 28 June 2018
- Cassan, A.; Kubas, D.; Beaulieu, J. P.; Dominik, M; et al. (2012). "One or more bound planets per Milky Way star from microlensing observations". Nature. 481 (7380): 167–169. arXiv:1202.0903. Bibcode:2012Natur.481..167C. doi:10.1038/nature10684. PMID 22237108

- Dressler, A. (March 1980). "Galaxy morphology in rich clusters - Implications for the formation and evolution of galaxies". The Astrophysical Journal. 236: 351–365. Bibcode:1980ApJ...236..351D. doi:10.1086/157753
- Planetary-astronomy, planetary-science-3: planetary-science.org, Retrieved 22 May 2018
- Emspak, Jesse. "Kepler Finds Bizarre Systems". International Business Times. International Business Times Inc. Retrieved 2 March 2011

6

Planet and its Types

A planet is a celestial body orbiting a star. Planets can be of different types, such as terrestrial planets, gas giants, ice giants, exoplanets and dwarf planets, classified on the basis of size, composition or distance from the Sun. This chapter explores these varied types of planets and their related attributes.

The International Astronomical Union (IAU) defines "planet" as a celestial body that, within the solar system,

(a) is in orbit around the sun;

(b) has sufficient mass for its self-gravity to overcome rigid body forces so that it assumes a hydrostatic equilibrium (near spherical) shape; and

(c) has "cleared the neighbourhood" around its orbit;

or within another system,

(1) is in orbit around a star or stellar remnants;

(2) has a mass below the limiting mass for thermonuclear fusion of deuterium; and

(3) is above the minimum mass/size requirement for planetary status in the Solar System.

Our solar system is thus considered to have eight planets: Mercury, Venus, Earth, Mars, Jupiter, Saturn, Uranus, and Neptune. Pursuant to a separate resolution, the IAU has designated three dwarf planets: Ceres, Pluto, and Eris. There have been more than two hundred planets discovered orbiting other stars to date.

Historically, there had been no formal scientific definition of "planet" and without one, the solar system had been considered to have various planets over the years. This changed when a resolution covering planets within our solar system was formally adopted by the IAU in 2006, limiting the number to eight. However, the IAU's position on those in other systems remains only a working definition in place since 2003, and as such, is easily subject to change. The IAU has not yet taken a position on free-floating objects of planetary mass outside star systems, other than to exclude those in young star clusters.

In ancient times, astronomers noted how certain lights moved across the sky in relation to the other stars. These objects were believed to orbit Earth, which was considered to be stationary. The "wandering" lights were first called "πλανήτης" (*planētēs*) by the Ancient Greeks, a Greek term meaning "wanderer," and it is from this that the word "planet" was derived.

In near-universal practice in the Western world, the planets in the solar system are named after Graeco-Roman gods, as, in Europe, it was the Greeks who named them, although the practice of naming planets after gods originated in the West with the Sumerians, who lived in modern-day Iraq in about 3000 B.C.E. Successive Mesopotamian civilizations, such as the Babylonians, retained the Sumerian naming convention but adapted it to their own very different pantheons. The Greeks borrowed much of their astronomy, including constellations and the zodiac, from the Babylonians, and by 600 B.C.E. had already begun using Babylonian concepts in their calculations.

The gods of Olympus, after whom the solar system's planets were named

The Greeks grafted the names of their own gods onto the Babylonian planet list, although there was some confusion in translation: for instance, the Babylonian Nergal was a god of war, and the Greeks, seeing this aspect of Nergal's persona, identified him with Ares, their god of war. However, Nergal, unlike Ares, was also a god of the dead and a god of pestilence.

Because of the influence of the Roman Empire and, later, the Roman Catholic Church, in most countries in the West the planets are known by their Roman (or Latin) names rather than the Greek. The Romans, who, like the Greeks, were Indo-Europeans, shared with them a common pantheon under different names but lacked the rich narrative traditions that Greek poetic culture had given their gods. During the later period of the Roman Republic, Roman writers borrowed much of the Greek narratives and applied them to their own pantheon, to the point where they became virtually indistinguishable. When the Romans studied Greek astronomy, they gave the planets their own gods' names.

In ancient times, there were five known planets; each presumed to be circling Earth according to the complex laws laid out by Ptolemy in the second century. They were, in the increasing order from the Earth: Mercury (called Hermes by the Greeks), Venus (Aphrodite), Mars (Ares), Jupiter (Zeus), and Saturn (Kronos). Although strictly the term "planetes" referred only to those five objects, the term was often expanded to include the sun and the moon.

Eventually, the Earth was added to the list of planets and the sun and moon were dropped in accordance with the heliocentric model. However, when subsequent planets were discovered in the eighteenth and nineteenth centuries, the naming practice was retained: Uranus (Ouranos) and Neptune (Poseidon). The Greeks still use their original names for the planets.

Some Romans, following a belief imported from Mesopotamia into Hellenistic Egypt, believed that the seven gods after whom the planets were named took hourly shifts in looking after affairs on Earth, in Ptolemaic orbit order listed inwards. As a result, a list of which god has charge of the first hour in each day came out as sun, moon, Mars, Mercury, Jupiter, Venus, Saturn, i.e. the usual weekday name order. Sunday, Monday, and Saturday are straightforward translations of these Roman names. In English the other days were renamed after Tiw, Wóden, Thunor, and Fríge, gods considered similar or equivalent to Mars, Mercury, Jupiter, and Venus respectively.

Since Earth was only generally accepted as a planet in the seventeenth century, there is no tradition of naming it after a god. Many of the Romance languages (including French, Italian, Spanish and Portuguese), which are descended from Latin, retain the old Roman name of *Terra* or some variation thereof. However, the non-Romance languages use their own respective native words. Again, the Greeks retain their original name, *Γή* (*Ge* or *Yi*); the Germanic languages, including English, use a variation of an ancient Germanic word *ertho*, "ground," as can be seen in the English *Earth*, the German *Erde*, the Dutch *Aarde*, and the Scandinavian *Jorde*. The same is true for the Sun and the Moon, though they are no longer considered planets.

Some non-European cultures use their own planetary naming systems. China and the countries of eastern Asia subject to Chinese cultural influence, such as Japan, Korea and Vietnam, use a naming system based on the five Chinese elements.

Former Planets

In ancient times, astronomers accepted as "planets" the seven visible objects that moved across the starfield: the Sun, the Moon, Mercury, Venus, Mars, Jupiter, and Saturn. Since then, many objects have qualified as planets for a time:

Body	Period of planethood	Solar system region	Present status	Notes
Sun	Antiquity to 1600s	Centre	Star	Planet under the geocentric model.
Moon	Antiquity to 1600s	Earth's orbit	Satellite	Planet under the geocentric model.
Ceres	1801-1864	Asteroid belt	Dwarf planet	Asteroid until at least 2006.
Pallas	1802-1864	Asteroid belt	Asteroid	
Juno	1804-1864	Asteroid belt	Asteroid	
Vesta	1807-1864	Asteroid belt	Asteroid	
Pluto	1930-2006	Kuiper belt	Dwarf planet	Officially accepted by IAU for this period.

Definition and Disputes

With the discovery during the latter half of the twentieth century of more objects within the solar system and large objects around other stars, dispute arose over what should constitute a planet.

There was particular disagreement over whether round objects that existed in belts, and large deuterium fusing objects should qualify.

In 2003, The International Astronomical Union (IAU) Working Group on Extrasolar Planets made a position statement on the definition of a planet that incorporated a working definition:

> *1) Objects with true masses below the limiting mass for thermonuclear fusion of deuterium (currently calculated to be 13 Jupiter masses for objects of solar metallicity) that orbit stars or stellar remnants are "planets" (no matter how they formed). The minimum mass/size required for an extrasolar object to be considered a planet should be the same as that used in our Solar System.*
>
> *2) Substellar objects with true masses above the limiting mass for thermonuclear fusion of deuterium are "brown dwarfs," no matter how they formed nor where they are located.*
>
> *3) Free-floating objects in young star clusters with masses below the limiting mass for thermonuclear fusion of deuterium are not "planets," but are "sub-brown dwarfs" (or whatever name is most appropriate).*

This definition has been widely used by astronomers when publishing discoveries in journals since this time, although it remains a temporary, working definition until a more permanent one is formally adopted. It also did not address the controversy over the lower mass limit.

However, in 2006, the general assembly of the IAU voted to pass a resolution that redefined planets within the solar system as:

> *A celestial body that is (a) in orbit around the Sun, (b) has sufficient mass for its self-gravity to overcome rigid body forces so that it assumes a hydrostatic equilibrium (nearly round) shape, and (c) has cleared the neighbourhood around its orbit.*

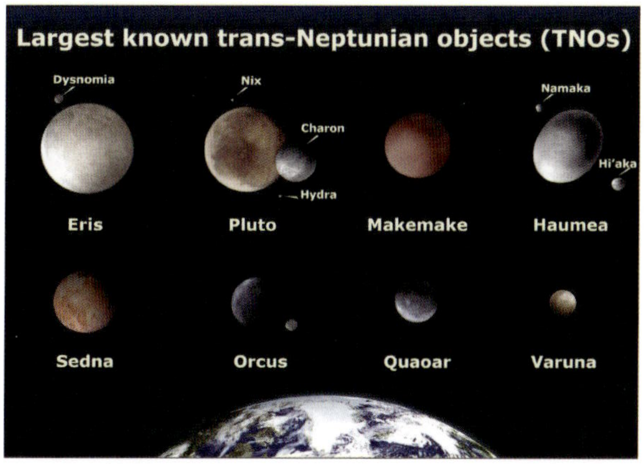

The largest Trans-Neptunian objects that prompted the IAU's decision

Under this definition, the solar system is considered to have eight planets. Bodies which fulfill the first two conditions but not the third (such as Pluto and Eris) are classified as dwarf planets, providing they are not also natural satellites of other planets. Originally an IAU committee had proposed a definition that would have included a much larger number of planets as it did not include (c) as a criterion. After much discussion, it was decided via a vote that those bodies should instead be classified as dwarf planets.

This definition is based in modern theories of planetary formation, in which planetary embryos initially clear their orbital neighbourhood of other smaller objects. As described by astronomer Steven Soter:

> *The end product of secondary disk accretion is a small number of relatively large bodies (planets) in either non-intersecting or resonant orbits, which prevent collisions between them. Asteroids and comets, including KBOs, differ from planets in that they can collide with each other and with planets.*

In the aftermath of the IAU's 2006 vote, there was criticism of the new definition. Part of the dispute centres around the belief that point (c) (clearing its orbit) should not have been listed, and that those objects now categorized as dwarf planets should actually be part of a broader planetary definition.

Beyond the scientific community, Pluto has held a strong cultural significance for many in the general public considering its planetary status during most of the twentieth century, in a similar way to Ceres and its kin in the 1800s. More recently, the discovery of Eris was widely reported in the media as the "tenth planet." The reclassification of all three objects as dwarf planets has attracted much media and public attention.

Formation

It is not known with certainty how planets are formed. The prevailing theory is that they are formed from those remnants of a nebula that do not condense under gravity to form a protostar. Instead, these remnants become a thin, protoplanetary disk of dust and gas revolving around the protostar and begin to condense about local concentrations of mass within the disc known as planetesimals. These concentrations become ever more dense until they collapse inward under gravity to form protoplanets. After a planet reaches a diameter larger than Earth's moon, it begins to accumulate an extended atmosphere. This serves to increase the capture rate of the planetesimals by a factor of ten.

When the protostar has grown such that it ignites to form a star, its solar wind blows away most of the disc's remaining material. Thereafter there still may be many protoplanets orbiting the star or each other, but over time many will collide, either to form a single larger planet or release material for other larger protoplanets or planets to absorb. Those objects that have become massive enough will capture most matter in their orbital neighbourhoods to become planets. Meanwhile, protoplanets that have avoided collisions may become natural satellites of planets through a process of gravitational capture, or remain in belts of other objects to become either dwarf planets or small solar system bodies.

The energetic impacts of the smaller planetesimals will heat up the growing planet, causing it to at least partially melt. The interior of the planet begins to differentiate by mass, developing a denser core. Smaller terrestrial planets lose most of their atmospheres because of this accretion, but the lost gases can be replaced by outgassing from the mantle and from the subsequent impact of comets. Smaller planets will lose any atmosphere they gain through various escape mechanisms.

With the discovery and observation of planetary systems around stars other than our own, it is becoming possible to elaborate, revise or even replace this account. The level of metallicity is now believed to determine the likelihood that a star will have planets. Hence it is thought less likely that a metal-poor, population II star will possess a more substantial planetary system than a metal-rich population I star.

Within the Solar System

The terrestrial planets: Mercury, Venus, Earth, Mars (sizes to scale)

The four gas giants against the Sun: Jupiter, Saturn, Uranus, Neptune (sizes to scale)

According to the IAU's current definitions, there are eight planets in the solar system. In increasing distance from the sun, they are:

(☿) Mercury, with no confirmed natural satellites

(♀) Venus, with no confirmed natural satellites

(⊕) Earth, with one confirmed natural satellite

(♂) Mars, with two confirmed natural satellites

(♃) Jupiter, with 63 confirmed natural satellites

(♄) Saturn, with 56 confirmed natural satellites

(⛢) Uranus, with 27 confirmed natural satellites

(♆) Neptune, with 13 confirmed natural satellites

The larger bodies of the solar system can be divided into categories based on their composition:

- Terrestrials: Planets (and possibly dwarf planets) that are similar to Earth with bodies largely composed of rock: Mercury, Venus, Earth, and Mars. If including dwarf planets, Ceres would also be counted, with as many as three other asteroids that might be added.
- Gas giants: Planets with a composition largely made up of gaseous material and are significantly more massive than terrestrials: Jupiter, Saturn, Uranus, and Neptune. Ice giants are a sub-class of gas giants, distinguished from gas giants by their depletion in hydrogen and helium, and a significant composition of rock and ice: Uranus and Neptune.
- Ice dwarfs: Objects that are composed mainly of ice, and do not have planetary mass. The dwarf planets Pluto and Eris are ice dwarfs, and several dwarf planetary candidates also qualify.

Attributes

All the planets revolve around the sun in the same direction counter-clockwise as seen from over the sun's north pole. The period of one revolution of a planet's orbit is known as its year. A planet's year depends on its distance from the sun. The farther a planet is from the sun, not only must it travel a longer distance but also its speed is slower, as it is less affected by the sun's gravity.

The planets also rotate around invisible axes through their centres. The period of one rotation of a planet is known as its day. All the planets rotate in a counter-clockwise direction, except for Venus, which rotates clockwise. There is great variation in the length of day between the planets, with Venus taking 243 Earth days to rotate, and the gas giants only a few hours.

Planets also have varying degrees of axial tilt; they lie at an angle to the plane of the sun's equator. This causes the amount of sunlight received by each hemisphere to vary over the course of its year; when the northern hemisphere points away from the sun, the southern hemisphere points towards it, and vice versa. Each planet therefore possesses seasons; changes to the climate over the course of its year. The point at which each hemisphere is farthest/nearest from the sun is known as its solstice. Each planet has two in the course of its orbit; when a planet's northern hemisphere has its summer solstice, when its day is longest, the southern has its winter solstice, when its day is shortest. Jupiter's axial tilt is very small, so its seasonal variation is minimal; Uranus, on the other hand, has an axial tilt so extreme it is virtually on its side, which means that its hemispheres are either perpetually in sunlight or perpetually in darkness around the time of its solstices.

All of the planets have atmospheres as their large masses mean gravity is strong enough to keep gaseous particles close to the surface. The larger gas giants are massive enough to keep large amounts of the light gases hydrogen and helium close by, although these gases mostly float into space around the smaller planets. Earth's atmosphere is greatly different to the other planets because of the various life processes that have transpired there, while the atmosphere of Mercury has mostly, although not entirely, been blasted away by the solar wind.

Many of the planets have natural satellites, called "moons," regardless of their size. The gas giants all have numerous moons in complex planetary systems. Many gas giant moons have similar features to the terrestrial planets and dwarf planets, and some have been studied for signs of life.

Planetary attributes											
	Name	Equatorial* diameter	Mass*	Orbital radius (AU)	Orbital period (years)	Inclination to Sun's equator (°)	Orbital eccentricity	Rotation period (days)	Moons	Rings	Atmosphere
Terrestrials	Mercury	0.39	0.06	0.39	0.24	3.38	0.206	58.64	none	no	minimal
	Venus	0.95	0.82	0.72	0.62	3.86	0.007	-243.02	none	no	CO_2, N_2
	Earth**	1.00	1.00	1.00	1.00	7.25	0.017	1.00	1	no	N_2, O_2
	Mars	0.53	0.11	1.52	1.88	5.65	0.093	1.03	2	no	CO_2, N_2
Gas giants	Jupiter	11.21	317.8	5.20	11.86	6.09	0.048	0.41	63	yes	H_2, He
	Saturn	9.41	95.2	9.54	29.46	5.51	0.054	0.43	56	yes	H_2, He
	Uranus	3.98	14.6	19.22	84.01	6.48	0.047	-0.72	27	yes	H_2, He
	Neptune	3.81	17.2	30.06	164.8	6.43	0.009	0.67	13	yes	H_2, He

*Measured relative to Earth. **

Dwarf Planets

Before the August 2006 decision, several objects were proposed by astronomers including at one stage by the IAU as planets. However, in 2006, several of these objects were reclassified as dwarf planets, objects distinct from planets. The IAU recognizes three dwarf planets in the solar system: Ceres, Pluto and Eris. Several other objects in both the asteroid belt and the Kuiper belt are under consideration, with as many as 50 that could eventually qualify. There may be as many as two hundred that could be discovered once the Kuiper Belt has been fully explored. Dwarf planets share many of the same characteristics as planets, although notable differences remain namely that they are not dominant in their orbits. Their attributes are:

Dwarf planetary attributes											
	Name	Equatorial* diameter	Mass*	Orbital radius (AU)	Orbital period (years)	Inclination to ecliptic (°)	Orbital eccentricity	Rotation period (days)	Moons	Rings	Atmosphere
Terrestrials	Ceres	0.08	0.0002	2.76	4.60	10.59	0.080	0.38	none	no	none
Ice dwarfs	Pluto	0.18	0.0022	39.48	248.09	17.14	0.249	-6.39	3	no	temporary
	Eris	0.19	0.0025	67.67	~557	44.19	0.442	~0.3	1	no	temporary

*Measured relative to the Earth.

By definition, all dwarf planets are members of larger populations. Ceres is the largest body in the asteroid belt, while Pluto is a member of the Kuiper belt and Eris is a member of the scattered disc. According to Mike Brown there may soon be over forty trans-Neptunian objects that qualify as dwarf planets under the IAU's recent definition.

Beyond the Solar System

Extrasolar Planets

Of the 209 extrasolar planets (those outside the solar system) discovered as of November 2006, most have masses which are about the same as, or larger than, Jupiter.

Exceptions include a number of planets discovered orbiting burned-out star remnants called pulsars, such as PSR B1257+12, the planets orbiting the stars Mu Arae, 55 Cancri and GJ 436, which are approximately Neptune-sized, and a planet orbiting Gliese 876 that is estimated to be about six to eight times as massive as Earth and is probably rocky in composition.

It is far from clear if the newly discovered large planets would resemble the gas giants in the solar system or if they are of an entirely different type as yet unknown, like ammonia giants or carbon planets. In particular, some of the newly discovered planets, known as "hot Jupiters," orbit extremely close to their parent stars, in nearly circular orbits. They therefore receive much more stellar radiation than the gas giants in the solar system, which makes it questionable whether they are the same type of planet at all. There is also a class of "hot Jupiters" that orbit so close to their star that their atmospheres are slowly blown away in a comet-like tail: the Chthonian planets.

Several projects have been proposed to create an array of space telescopes to search for extrasolar planets with masses comparable to the Earth. The NASA Terrestrial Planet Finder was one such program, but as of February 2006 this program had been put on indefinite hold. The European Space Agency is considering a comparable mission called "Darwin." The frequency of occurrence of such terrestrial planets is one of the variables in the Drake equation which estimates the number of intelligent, communicating civilizations that exist in our galaxy.

In 2005, astronomers detected a planet in a triple star system, a finding that challenges current theories of planetary formation. The planet, a gas giant slightly larger than Jupiter, orbits the main star of the HD 188753 system, in the constellation Cygnus, and is hence known as HD 188753 Ab. The stellar trio (yellow, orange, and red) is about 149 light years from Earth. The planet, which is at least 14 percent larger than Jupiter, orbits the main star (HD 188753 A) once every 80 hours or so (3.3 days), at a distance of about 8 Gm, a twentieth of the distance between Earth and the sun. The other two stars whirl tightly around each other in 156 days, and circle the main star every 25.7 years at a distance from the main star that would put them between Saturn and Uranus in the solar system. The latter stars invalidate the leading hot Jupiter formation theory, which holds that these planets form at "normal" distances and then migrate inward through some debatable mechanism. This could not have occurred here; the outer star pair would have disrupted outer planet formation.

Interstellar "Planets"

Several computer simulations of stellar and planetary system formation have suggested that some objects of planetary mass would be ejected into interstellar space. Some scientists have argued that such objects found roaming in deep space should be classed as "planets." However, many others argue that only planemos that directly orbit stars should qualify as planets, preferring to use the terms "planetary body," "planetary mass object" or "planemo" for similar free-floating objects (as well as planet-size moons). The IAU's working definition on extrasolar planets takes no position on the issue. The discoverers of the bodies mentioned above decided to avoid the debate over what constitutes a planet by referring to the objects as planemos. However, the original IAU proposal for the 2006 definition of planet favored the star-orbiting criterion, although the final draft avoided the issue.

Exoplanets

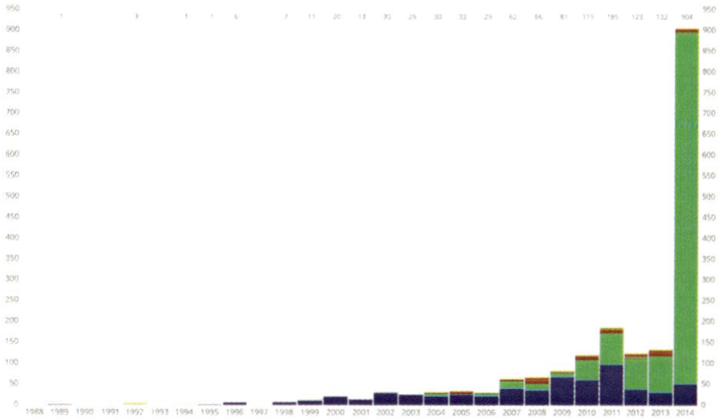

Exoplanets, by year of discovery

An exoplanet (extrasolar planet) is a planet outside the Solar System, there are 3,815 confirmed planets in 2,853 systems, with 633 systems having more than one planet.

In early 1992, radio astronomers Aleksander Wolszczan and Dale Frail announced the discovery of two planets orbiting the pulsar PSR 1257+12. This discovery was confirmed, and is generally considered to be the first definitive detection of exoplanets. These pulsar planets are believed to have formed from the unusual remnants of the supernova that produced the pulsar, in a second round of planet formation, or else to be the remaining rocky cores of giant planets that survived the supernova and then decayed into their current orbits.

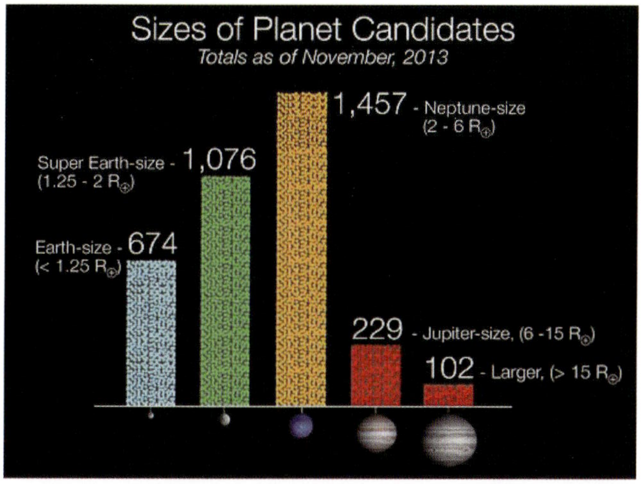

Sizes of *Kepler* Planet Candidates – based on 2,740 candidates orbiting 2,036 stars

The first confirmed discovery of an extrasolar planet orbiting an ordinary main-sequence star occurred on 6 October 1995, when Michel Mayor and Didier Queloz of the University of Geneva announced the detection of an exoplanet around 51 Pegasi. From then until the Kepler mission most known extrasolar planets were gas giants comparable in mass to Jupiter or larger as they were more easily detected. The catalog of Kepler candidate planets consists mostly of planets the size of Neptune and smaller, down to smaller than Mercury.

There are types of planets that do not exist in the Solar System: super-Earths and mini-Neptunes, which could be rocky like Earth or a mixture of volatiles and gas like Neptune a radius of 1.75 times that of Earth is a possible dividing line between the two types of planet. There are hot Jupiters that orbit very close to their star and may evaporate to become chthonian planets, which are the left-over cores. Another possible type of planet is carbon planets, which form in systems with a higher proportion of carbon than in the Solar System.

A 2012 study, analyzing gravitational microlensing data, estimates an average of at least 1.6 bound planets for every star in the Milky Way.

On December 20, 2011, the Kepler Space Telescope team reported the discovery of the first Earth-size exoplanets, Kepler-20e and Kepler-20f, orbiting a Sun-like star, Kepler-20.

Around 1 in 5 Sun-like stars have an "Earth-sized" planet in the habitable zone, so the nearest would be expected to be within 12 light-years distance from Earth. The frequency of occurrence of such terrestrial planets is one of the variables in the Drake equation, which estimates the number of intelligent, communicating civilizations that exist in the Milky Way.

There are exoplanets that are much closer to their parent star than any planet in the Solar System is to the Sun, and there are also exoplanets that are much farther from their star. Mercury, the closest planet to the Sun at 0.4 AU, takes 88 days for an orbit, but the shortest known orbits for exoplanets take only a few hours, e.g. Kepler-70b. The Kepler-11 system has five of its planets in shorter orbits than Mercury's, all of them much more massive than Mercury. Neptune is 30 AU from the Sun and takes 165 years to orbit, but there are exoplanets that are hundreds of AU from their star and take more than a thousand years to orbit, e.g. 1RXS1609 b.

The next few space telescopes to study exoplanets are expected to be Gaia launched in December 2013, CHEOPS in 2018, TESS in 2018, and the James Webb Space Telescope in 2019.

Planetary-mass Objects

Artist's impression of a super-Jupiter around the brown dwarf 2M1207

A planetary-mass object (PMO), planemo, or planetary body is a celestial object with a mass that falls within the range of the definition of a planet: massive enough to achieve hydrostatic equilibrium (to be rounded under its own gravity), but not enough to sustain core fusion like a star. By definition, all planets are *planetary-mass objects*, but the purpose of this term is to refer to objects that do not conform to typical expectations for a planet. These include dwarf planets, which are rounded by their own gravity but not massive enough to clear their own orbit, the larger moons, and free-floating planemos, which may have been ejected from a system (rogue planets) or formed through cloud-collapse rather than accretion (sometimes called sub-brown dwarfs).

Rogue Planets

Several computer simulations of stellar and planetary system formation have suggested that some objects of planetary mass would be ejected into interstellar space. Some scientists have argued that such objects found roaming in deep space should be classed as "planets", although others have suggested that they should be called low-mass brown dwarfs.

Sub-brown Dwarfs

Stars form via the gravitational collapse of gas clouds, but smaller objects can also form via cloud-collapse. Planetary-mass objects formed this way are sometimes called sub-brown dwarfs. Sub-brown dwarfs may be free-floating such as Cha 110913-773444 and OTS 44, or orbiting a larger object such as 2MASS J04414489+2301513.

Binary systems of sub-brown dwarfs are theoretically possible; Oph 162225-240515 was initially thought to be a binary system of a brown dwarf of 14 Jupiter masses and a sub-brown dwarf of 7 Jupiter masses, but further observations revised the estimated masses upwards to greater than 13 Jupiter masses, making them brown dwarfs according to the IAU working definitions.

Former Stars

In close binary star systems one of the stars can lose mass to a heavier companion. Accretion-powered pulsars may drive mass loss. The shrinking star can then become a planetary-mass object. An example is a Jupiter-mass object orbiting the pulsar PSR J1719-1438. These shrunken white dwarfs may become a helium planet or carbon planet.

Satellite Planets and Belt Planets

Some large satellites are of similar size or larger than the planet Mercury, e.g. Jupiter's Galilean moons and Titan. Alan Stern has argued that location should not matter and that only geophysical attributes should be taken into account in the definition of a planet, and proposes the term *satellite planet* for a planet-sized satellite. Likewise, dwarf planets in the asteroid belt and Kuiper belt should be considered planets according to Stern.

Captured Planets

Free-floating planets in stellar clusters have similar velocities to the stars and so can be recaptured. They are typically captured into wide orbits between 100 and 10^5 AU. The capture efficiency decreases with increasing cluster volume, and for a given cluster size it increases with the host/primary mass. It is almost independent of the planetary mass. Single and multiple planets could be captured into arbitrary unaligned orbits, non-coplanar with each other or with the stellar host spin, or pre-existing planetary system.

Physical Characteristics

Mass

A planet's defining physical characteristic is that it is massive enough for the force of its own gravity to dominate over the electromagnetic forces binding its physical structure, leading to a state of hydrostatic equilibrium. This effectively means that all planets are spherical or spheroidal. Up to a certain mass, an object can be irregular in shape, but beyond that point, which varies depending on the chemical makeup of the object, gravity begins to pull an object towards its own centre of mass until the object collapses into a sphere.

Mass is also the prime attribute by which planets are distinguished from stars. The upper mass limit for planethood is roughly 13 times Jupiter's mass for objects with solar-type isotopic abundance, beyond which it achieves conditions suitable for nuclear fusion. Other than the Sun, no objects of such mass exist in the Solar System; but there are exoplanets of this size. The 13-Jupiter-mass limit is not universally agreed upon and the Extrasolar Planets Encyclopaedia includes objects up to 20 Jupiter masses, and the Exoplanet Data Explorer up to 24 Jupiter masses.

The smallest known planet is PSR B1257+12A, one of the first extrasolar planets discovered, which was found in 1992 in orbit around a pulsar. Its mass is roughly half that of the planet Mercury. The smallest known planet orbiting a main-sequence star other than the Sun is Kepler-37b, with a mass (and radius) slightly higher than that of the Moon.

Internal Differentiation

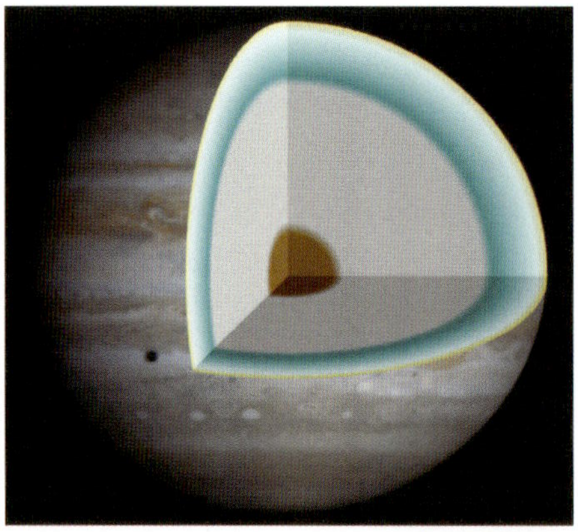

Illustration of the interior of Jupiter, with a rocky core overlaid by a deep layer of metallic hydrogen

Every planet began its existence in an entirely fluid state; in early formation, the denser, heavier materials sank to the centre, leaving the lighter materials near the surface. Each therefore has a differentiated interior consisting of a dense planetary core surrounded by a mantle that either is or was a fluid. The terrestrial planets are sealed within hard crusts, but in the giant planets the mantle simply blends into the upper cloud layers. The terrestrial planets have cores of elements such as iron and nickel, and mantles of silicates. Jupiter and Saturn are believed to have cores of rock and metal surrounded by mantles of metallic hydrogen. Uranus and Neptune, which are smaller, have rocky cores surrounded by mantles of water, ammonia, methane and other ices. The fluid action within these planets' cores creates a geodynamo that generates a magnetic field.

Atmosphere

Earth's atmosphere

All of the Solar System planets except Mercury have substantial atmospheres because their gravity is strong enough to keep gases close to the surface. The larger giant planets are massive enough to keep large amounts of the light gases hydrogen and helium, whereas the smaller planets lose these gases into space. The composition of Earth's atmosphere is different from the other planets because the various life processes that have transpired on the planet have introduced free molecular oxygen.

Planetary atmospheres are affected by the varying insolation or internal energy, leading to the formation of dynamic weather systems such as hurricanes, (on Earth), planet-wide dust storms (on Mars), a greater-than-Earth-sized anticyclone on Jupiter (called the Great Red Spot), and holes in the atmosphere (on Neptune). At least one extrasolar planet, HD 189733 b, has been claimed to have such a weather system, similar to the Great Red Spot but twice as large.

Hot Jupiters, due to their extreme proximities to their host stars, have been shown to be losing their atmospheres into space due to stellar radiation, much like the tails of comets. These planets may have vast differences in temperature between their day and night sides that produce supersonic winds, although the day and night sides of HD 189733 b appear to have very similar temperatures, indicating that that planet's atmosphere effectively redistributes the star's energy around the planet.

Magnetosphere

One important characteristic of the planets is their intrinsic magnetic moments, which in turn give rise to magnetospheres. The presence of a magnetic field indicates that the planet is still geologically alive. In other words, magnetized planets have flows of electrically conducting material in their interiors, which generate their magnetic fields. These fields significantly change the interaction of the planet and solar wind. A magnetized planet creates a cavity in the solar wind around itself called the magnetosphere, which the wind cannot penetrate. The magnetosphere can be much larger than the planet itself. In contrast, non-magnetized planets have only small magnetospheres induced by interaction of the ionosphere with the solar wind, which cannot effectively protect the planet.

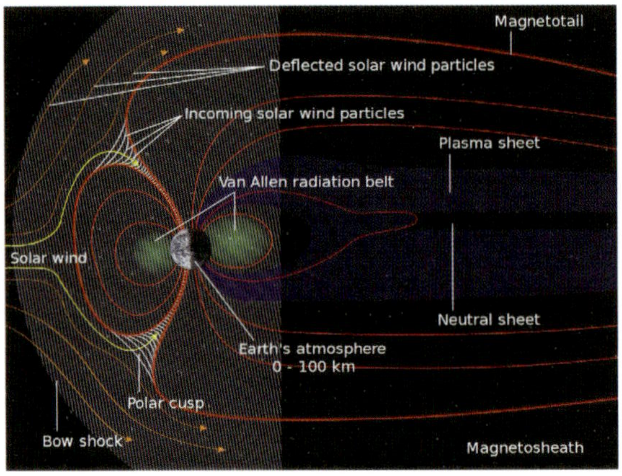

Earth's magnetosphere (diagram)

Of the eight planets in the Solar System, only Venus and Mars lack such a magnetic field. In addition, the moon of Jupiter Ganymede also has one. Of the magnetized planets the magnetic field of Mercury is the weakest, and is barely able to deflect the solar wind. Ganymede's magnetic field is several times larger, and Jupiter's is the strongest in the Solar System (so strong in fact that it poses a serious health risk to future manned missions to its moons). The magnetic fields of the other giant planets are roughly similar in strength to that of Earth, but their magnetic moments are significantly larger. The magnetic fields of Uranus and Neptune are strongly tilted relative the rotational axis and displaced from the centre of the planet.

In 2004, a team of astronomers in Hawaii observed an extrasolar planet around the star HD 179949, which appeared to be creating a sunspot on the surface of its parent star. The team hypothesized that the planet's magnetosphere was transferring energy onto the star's surface, increasing its already high 7,760 °C temperature by an additional 400 °C.

Secondary Characteristics

The rings of Saturn

Several planets or dwarf planets in the Solar System (such as Neptune and Pluto) have orbital periods that are in resonance with each other or with smaller bodies (this is also common in satellite systems). All except Mercury and Venus have natural satellites, often called "moons". Earth has one, Mars has two, and the giant planets have numerous moons in complex planetary-type systems. Many moons of the giant planets have features similar to those on the terrestrial planets and dwarf planets, and some have been studied as possible abodes of life (especially Europa).

The four giant planets are also orbited by planetary rings of varying size and complexity. The rings are composed primarily of dust or particulate matter, but can host tiny 'moonlets' whose gravity shapes and maintains their structure. Although the origins of planetary rings is not precisely known, they are believed to be the result of natural satellites that fell below their parent planet's Roche limit and were torn apart by tidal forces.

No secondary characteristics have been observed around extrasolar planets. The sub-brown dwarf Cha 110913-773444, which has been described as a rogue planet, is believed to be orbited by a tiny protoplanetary disc and the sub-brown dwarf OTS 44 was shown to be surrounded by a substantial protoplanetary disk of at least 10 Earth masses.

Terrestrial Planet

The term terrestrial is derived from the Latin word 'terra', which means earth. Thus, terrestrial planets can be described as planets that are earth-like in nature, meaning those that are similar in composition and structure to that of the earth. They are also known as 'telluric planets' or 'rocky planets'.

Terrestrial planets usually have a metallic core - consisting mostly of iron - and a silicate mantle. They have topological features like mountains, valleys, craters, etc. on their surface. Unlike giant planets (which have a primary atmosphere), terrestrial planets have a secondary atmosphere. Terrestrial planets are known to have few or no satellites of their own. For example: Earth has only one satellite, Mars has two, whereas Mercury and Venus have none.

Composition and Characteristics

All terrestrial planets have approximately the same type of structure: a central metallic core composed of mostly iron, with a surrounding silicate mantle. Such planets have common surface features, which include canyons, craters, mountains, volcanoes, and other similar structures, depending on the presence of water and tectonic activity.

Terrestrial planets also have secondary atmospheres, which are generated through volcanism or comet impacts. This also differentiates them from gas giants, where the planetary atmospheres are primary and were captured directly from the original solar nebula.

Terrestrial planets are also known for having few or no moons. Venus and Mercury have no moons, while Earth has only the one (the Moon). Mars has two satellites, Phobos and Deimos, but these are more akin to large asteroids than actual moons. Unlike the gas giants, terrestrial planets also have no planetary ring systems.

The Earth's interior structure, shown here as consisting of multiple "layers"

Solar Terrestrial Planets

All those planets found within the Inner Solar System – Mercury, Venus, Earth and Mars – are examples of terrestrial planets. Each are composed primarily of silicate rock and metal, which is differentiated between a dense, metallic core and a silicate mantle. The Moon is similar, but has a much smaller iron core.

Io and Europa are also satellites that have internal structures similar to that of terrestrial planets. In the case of the former, models of the moon's composition suggest that the mantle is composed primarily of silicate rock and iron, which surrounds a core of iron and iron sulphide. Europa, on the other hand, is believed to have an iron core that is surrounded by an outer layer of water.

Dwarf planets, like Ceres and Pluto, and other large asteroids are similar to terrestrial planets in the fact that they do have a solid surface. However, they differ in that they are, on average, composed of more icy materials than rock.

Extrasolar Terrestrial Planets

Most of the planets detected outside of the Solar System have been gas giants, owing to the fact that they are easier to spot. However, since 2005, hundreds of potentially terrestrial extrasolar planets have been found – mainly by the Kepler space mission. Most of these have been what is known as "super-Earths" (i.e. planets with masses between Earth's and Neptune's).

Examples of extrasolar terrestrial planets include Gliese 876 d, a planet that has a mass 7 to 9 times that of Earth. This planet orbits the red dwarf Gliese 876, which is located approximately 15 light years from Earth. The existence of three (or possibly four) terrestrial exoplanets was also confirmed between 2007 and 2010 in the Gliese 581 system, another red dwarf roughly 20 light years from Earth.

The smallest of these, Gliese 581 e, is only about 1.9 Earth masses, but orbits very close to the star. Two others, Gliese 581 c and Gliese 581 d, as well as a proposed fourth planet (Gliese 581 g) are more-massive super-Earths orbiting in or close to the habitable zone of the star. If true, this could mean that these worlds are potentially habitable Earth-like planets.

The first confirmed terrestrial exoplanet, Kepler-10b – a planet with between 3 and 4 Earth masses and located some 460 light years from Earth – was found in 2011 by the Kepler space mission. In that same year, the Kepler Space Observatory team released a list of 1235 extrasolar planet candidates, including six that were "Earth-size" or "super-Earth-size" (i.e. less than 2 Earth radii) and which were located within their stars' habitable zones.

Since then, Kepler has discovered hundreds of planets ranging from Moon-sized to super-Earths, with many more candidates in this size range. As of January, 2013, 2740 planet candidates have been discovered.

Categories

Scientists have proposed several categories for classifying terrestrial planets. Silicate planets are the standard type of terrestrial planet seen in the Solar System, which are composed primarily of a silicon-based rocky mantle and a metallic (iron) core.

Iron planets are a theoretical type of terrestrial planet that consists almost entirely of iron and therefore has a greater density and a smaller radius than other terrestrial planets of comparable mass. Planets of this type are believed to form in the high-temperature regions close to a star, and where the protoplanetary disk is rich in iron. Mercury is possible example, which formed close to our Sun and has a metallic core equal to 60–70% of its planetary mass.

Coreless planets are another theoretical type of terrestrial planet, one that consists of silicate rock but has no metallic core. In other words, coreless planets are the opposite of an iron planet. Coreless planets are believed to form farther from the star where volatile oxidizing material is more common. Though the Solar System has no coreless planets, chondrite asteroids and meteorites are common.

And then there are Carbon planets (aka. "diamond planets"), a theoretical class of planets that are composed of a metal core surrounded by primarily carbon-based minerals. Again, the Solar System has no planets that fit this description, but has an abundance of carbonaceous asteroids.

Non-terrestrial Planets

Not all planets are terrestrial. In our solar system, Jupiter, Saturn, Uranus and Neptune are gas giants, also known as jovian planets. It's unclear what the dividing line is between a rocky planet and a terrestrial planet; some super-Earths may have a liquid surface, for example. In our solar system, gas giants are much bigger than terrestrial planets, and they have thick atmospheres full of hydrogen and helium. On Jupiter and Saturn, hydrogen and helium make up most of the planet, while on Uranus and Neptune, the elements make up just the outer envelope. Landing on a gas giant is nearly impossible, since only the core is solid, and the atmosphere is storm-filled and very thick. It's also inhospitable to life as we know it.

Giant Planet

Basic Characteristics

The giant planets are very far from the Sun. Jupiter is more than five times farther from the Sun than Earth's distance (5 AU), and takes just under 12 years to circle the Sun. Saturn is about twice as far away as Jupiter (almost 10 AU) and takes nearly 30 years to complete one orbit. Uranus orbits at 19 AU with a period of 84 years, while Neptune, at 30 AU, requires 165 years for each circuit of the Sun. These long timescales make it difficult for us short-lived humans to study seasonal change on the outer planets.

Jupiter and Saturn have many similarities in composition and internal structure, although Jupiter is nearly four times more massive. Uranus and Neptune are smaller and differ in composition and internal structure from their large siblings. Some of the main properties of these four planets are summarized in table below.

Table: Basic Properties of the Jovian Planets

Planet	Distance (AU)	Period (years)	Diameter (km)	Mass (Earth = 1)	Density (g/cm^3)	Rotation (hours)
Jupiter	5.2	11.9	142,800	318	1.3	9.9
Saturn	9.5	29.5	120,540	95	0.7	10.7
Uranus	19.2	84.1	51,200	14	1.3	17.2
Neptune	30.0	164.8	49,500	17	1.6	16.1

Jupiter, the giant among giants, has enough mass to make 318 Earths. Its diameter is about 11 times that of Earth (and about one tenth that of the Sun). Jupiter's average density is 1.3 g/cm^3, much lower than that of any of the terrestrial planets. (Recall that water has a density of 1 g/cm^3.) Jupiter's material is spread out over a volume so large that more than 1400 Earths could fit within it.

Saturn's mass is 95 times that of Earth, and its average density is only 0.7 g/cm^3 the lowest of any planet. Since this is less than the density of water, Saturn would be light enough to float.

Uranus and Neptune each have a mass about 15 times that of Earth and, hence, are only 5% as massive as Jupiter. Their densities of 1.3 g/cm^3 and 1.6 g/cm^3, respectively, are much higher than that of Saturn. This is one piece of evidence that tells us that their composition must differ fundamentally from the gas giants. When astronomers began to discover other planetary systems (exoplanets), we found that planets the size of Uranus and Neptune are common, and that there are even more exoplanets intermediate in size between Earth and these ice giants, a type of planet not found in our solar system.

Appearance and Rotation

When we look at the planets, we see only their atmospheres, composed primarily of hydrogen and helium gas. The uppermost clouds of Jupiter and Saturn, the part we see when looking down at these planets from above, are composed of ammonia crystals. On Neptune, the upper clouds are made of methane. On Uranus, we see no obvious cloud layer at all, but only a deep and featureless haze.

Seen through a telescope, Jupiter is a colorful and dynamic planet. Distinct details in its cloud patterns allow us to determine the rotation rate of its atmosphere at the cloud level, although such atmosphere rotation may have little to do with the spin of the underlying planet. Much more fundamental is the rotation of the mantle and core; these can be determined by periodic variations in radio waves coming from Jupiter, which are controlled by its magnetic field. Since the magnetic field (which we will discuss below) originates deep inside the planet, it shares the rotation of the interior. The rotation period we measure in this way is 9 hours 56 minutes, which gives Jupiter the shortest "day" of any planet. In the same way, we can measure that the underlying rotation period of Saturn is 10 hours 40 minutes. Uranus and Neptune have slightly longer rotation periods of about 17 hours, also determined from the rotation of their magnetic fields.

Remember that Earth and Mars have seasons because their spin axes, instead of "standing up straight," are tilted relative to the orbital plane of the solar system. This means that as Earth revolves around the Sun, sometimes one hemisphere and sometimes the other "leans into" the Sun.

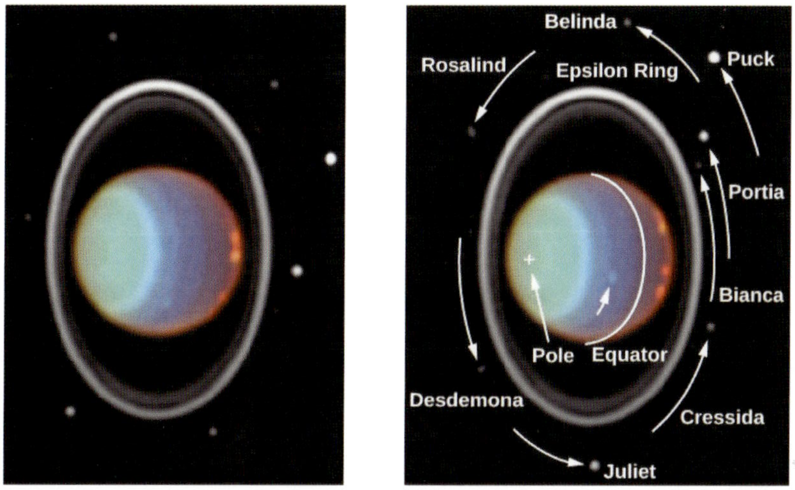

Figure: Uranus in the Infrared. The left-hand panel shows Uranus, with its rings, moons and some surface features. Due to the tilt of Uranus, we see the rings completely surround the planet, giving the appearance of a

What are the seasons like for the giant planets? The spin axis of Jupiter is tilted by only 3°, so there are no seasons to speak of. Saturn, however, does have seasons, since its spin axis is inclined at 27° to the perpendicular to its orbit. Neptune has about the same tilt as Saturn (29°); therefore, it experiences similar seasons (only more slowly). The strangest seasons of all are on Uranus, which has a spin axis tilted by 98° with respect to the north direction. Practically speaking, we can say that Uranus orbits on its side, and its ring and moon system follow along, orbiting about Uranus' equator.

We don't know what caused Uranus to be tipped over like this, but one possibility is a collision with a large planetary body when our system was first forming. Whatever the cause, this unusual tilt creates dramatic seasons. When Voyager 2 arrived at Uranus, its south pole was facing directly into the Sun. The southern hemisphere was experiencing a 21-year sunlit summer, while during that same period the northern hemisphere was plunged into darkness. For the next 21-year season, the Sun shines on Uranus' equator, and both hemispheres go through cycles of light and dark as the planet rotates. Then there are 21 years of an illuminated northern hemisphere and a dark southern hemisphere. After that the pattern of alternating day and night repeats.

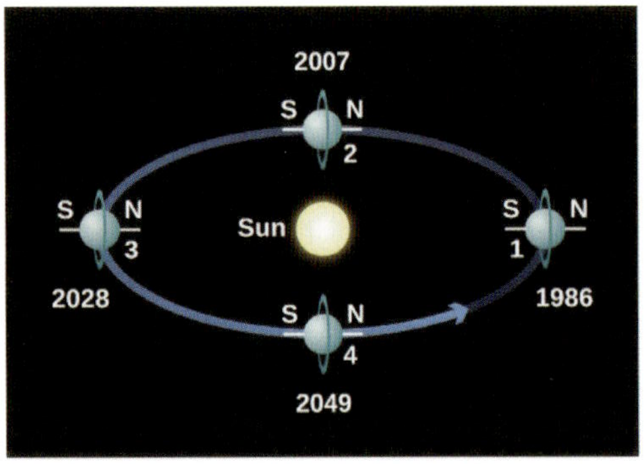

(a)

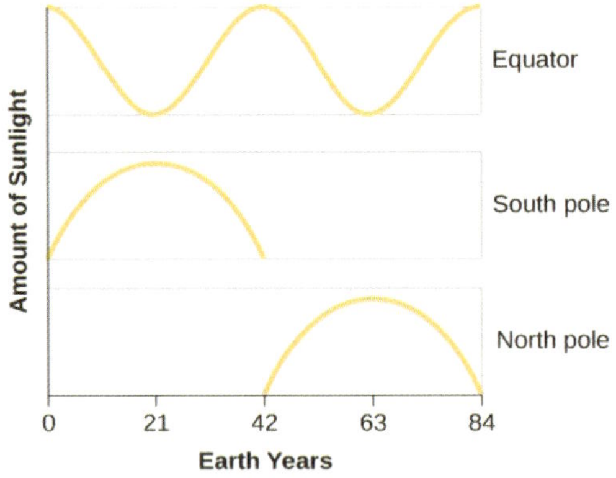

(b)

Figure: Strange Seasons on Uranus. (a) This diagram shows the orbit of Uranus as seen from above. (b) This graph compares the amount of sunlight seen at the poles and the equator of Uranus over the course of its 84-year revolution around the Sun

Just as on Earth, the seasons are even more extreme at the poles. If you were to install a floating platform at the south pole of Uranus, for example, it would experience 42 years of light and 42 years of darkness. Any future astronauts crazy enough to set up camp there could spend most of their lives without ever seeing the Sun.

Composition and Structure

Although we cannot see into these planets, astronomers are confident that the interiors of Jupiter and Saturn are composed primarily of hydrogen and helium. Of course, these gases have been measured only in their atmosphere, but calculations first carried out more than 50 years ago showed that these two light gases are the only possible materials out of which a planet with the observed masses and densities of Jupiter and Saturn could be constructed.

The deep internal structures of these two planets are difficult to predict. This is mainly because these planets are so big that the hydrogen and helium in their centres become tremendously compressed and behave in ways that these gases can never behave on Earth. The best theoretical models we have of Jupiter's structure predict a central pressure greater than 100 million bars and a central density of about 31 g/cm^3. (By contrast, Earth's core has a central pressure of 4 million bars and a central density of 17 g/cm^3.)

At the pressures inside the giant planets, familiar materials can take on strange forms. A few thousand kilometers below the visible clouds of Jupiter and Saturn, pressures become so great that hydrogen changes from a gaseous to a liquid state. Still deeper, this liquid hydrogen is further compressed and begins to act like a metal, something it never does on Earth. (In a metal, electrons are not firmly attached to their parent nuclei but can wander around. This is why metals are such good conductors of electricity.) On Jupiter, the greater part of the interior is liquid metallic hydrogen.

Because Saturn is less massive, it has only a small volume of metallic hydrogen, but most of its interior is liquid. Uranus and Neptune are too small to reach internal pressures sufficient to liquefy hydrogen. We will return to the discussion of the metallic hydrogen layers when we examine the magnetic fields of the giant planets.

Each of these planets has a core composed of heavier materials, as demonstrated by detailed analyses of their gravitational fields. Presumably these cores are the original rock-and-ice bodies that formed before the capture of gas from the surrounding nebula. The cores exist at pressures of tens of millions of bars. While scientists speak of the giant planet cores being composed of rock and ice, we can be sure that neither rock nor ice assumes any familiar forms at such pressures and temperatures. Remember that what is really meant by "rock" is any material made up primarily of iron, silicon, and oxygen, while the term "ice" in this chapter denotes materials composed primarily of the elements carbon, nitrogen, and oxygen in combination with hydrogen.

Figure illustrates the likely interior structures of the four jovian planets. It appears that all four have similar cores of rock and ice. On Jupiter and Saturn, the cores constitute only a few percent of the total mass, consistent with the initial composition of raw materials. However, most of the mass of Uranus and Neptune resides in these cores, demonstrating that the two outer planets were unable to attract massive quantities of hydrogen and helium when they were first forming.

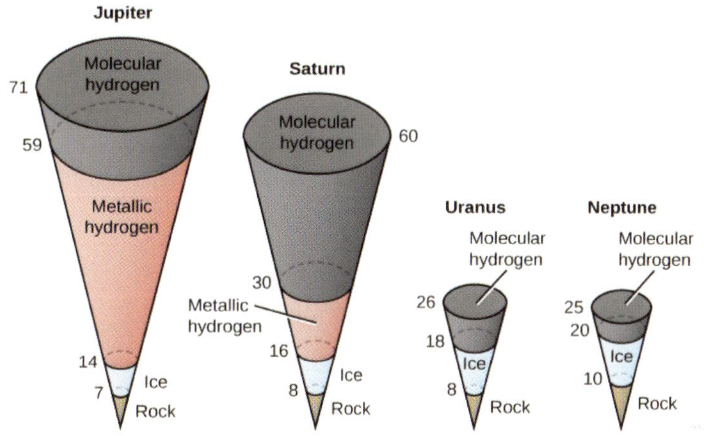

Figure: Internal Structures of the Jovian Planets. Jupiter and Saturn are composed primarily of hydrogen and helium (but hydrogen dominates), but Uranus and Neptune consist in large part of compounds of carbon, nitrogen, and oxygen. (The diagrams are drawn to scale; numbers show radii in thousands of kilometers.)

Internal Heat Sources

Because of their large sizes, all the giant planets were strongly heated during their formation by the collapse of surrounding material onto their cores. Jupiter, being the largest, was the hottest. Some of this primordial heat can still remain inside such large planets. In addition, it is possible for giant, largely gaseous planets to generate heat after formation by slowly contracting. (With so large a mass, even a minuscule amount of shrinking can generate significant heat.) The effect of these internal energy sources is to raise the temperatures in the interiors and atmospheres of the planets higher than we would expect from the heating effect of the Sun alone.

Jupiter has the largest internal energy source, amounting to 4×10^{17} watts; that is, it is heated from inside with energy equivalent to 4 million billion 100-watt lightbulbs. This energy is about the same as the total solar energy absorbed by Jupiter. The atmosphere of Jupiter is therefore something of a cross between a normal planetary atmosphere (like Earth's), which obtains most of its energy from the Sun, and the atmosphere of a star, which is entirely heated by an internal energy source. Most of the internal energy of Jupiter is primordial heat, left over from the formation of the planet 4.5 billon years ago.

Saturn has an internal energy source about half as large as that of Jupiter, which means (since its mass is only about one quarter as great) that it is producing twice as much energy per kilogram of material as does Jupiter. Since Saturn is expected to have much less primordial heat, there must be another source at work generating most of this 2×10^{17} watts of power. This source is the separation of helium from hydrogen in Saturn's interior. In the liquid hydrogen mantle, the heavier helium forms droplets that sink toward the core, releasing gravitational energy. In effect, Saturn is still differentiating letting lighter material rise and heavier material fall.

Uranus and Neptune are different. Neptune has a small internal energy source, while Uranus does not emit a measurable amount of internal heat. As a result, these two planets have almost the same atmospheric temperature, in spite of Neptune's greater distance from the Sun. No one knows why these two planets differ in their internal heat, but all this shows how nature can contrive to make each world a little bit different from its neighbours.

Magnetic Fields

Each of the giant planets has a strong magnetic field, generated by electric currents in its rapidly spinning interior. Associated with the magnetic fields are the planets' magnetospheres, which are regions around the planet within which the planet's own magnetic field dominates over the general interplanetary magnetic field. The magnetospheres of these planets are their largest features, extending millions of kilometers into space.

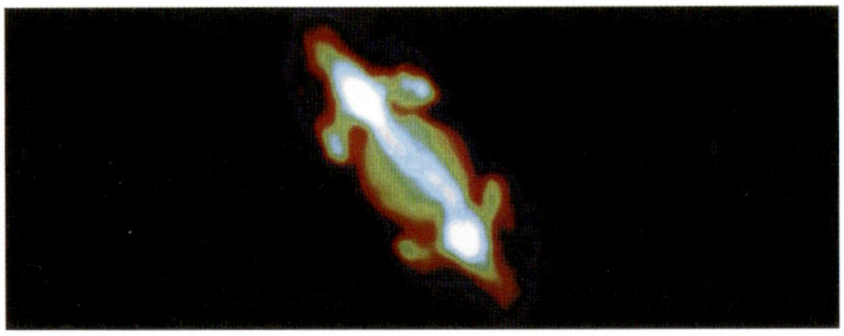

Figure: Jupiter in Radio Waves

This false-color image of Jupiter was made with the Very Large Array (of radio telescopes) in New Mexico. We see part of the magnetosphere, brightest in the middle because the largest number of charged particles are in the equatorial zone of Jupiter. The planet itself is slightly smaller than the green oval in the centre. Different colors are used to indicate different intensities of synchrotron radiation.

In the late 1950s, astronomers discovered that Jupiter was a source of radio waves that got more intense at longer rather than at shorter wavelengths just the reverse of what is expected from thermal radiation (radiation caused by the normal vibrations of particles within all matter). Such behavior is typical, however, of the radiation emitted when high-speed electrons are accelerated by a magnetic field. We call this synchrotron radiation because it was first observed on Earth in particle accelerators, called synchrotrons. This was our first hint that Jupiter must have a strong magnetic field.

Later observations showed that the radio waves are coming from a region surrounding Jupiter with a diameter several times that of the planet itself (Figure 4). The evidence suggested that a vast number of charged atomic particles must be circulating around Jupiter, spiraling around the lines of force of a magnetic field associated with the planet. This is just what we observe happening, but on a smaller scale, in the Van Allen belt around Earth. The magnetic fields of Saturn, Uranus, and Neptune, discovered by the spacecraft that first passed close to these planets, work in a similar way, but are not as strong.

Subtypes

Gas Giants

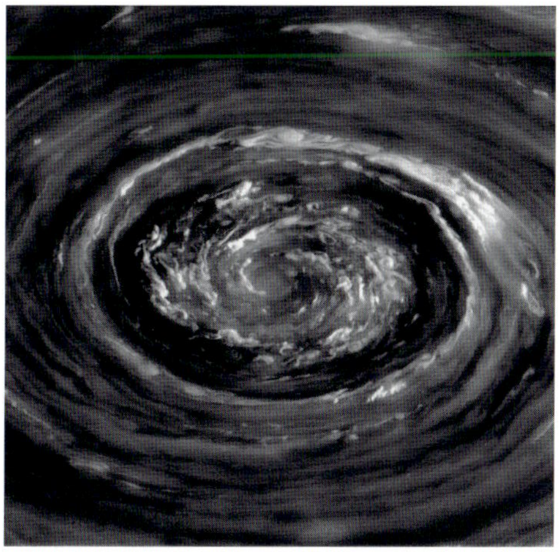

Saturn's north polar vortex

Gas giants consist mostly of hydrogen and helium. The Solar System's gas giants, Jupiter and Saturn, have heavier elements making up between 3 and 13 percent of their mass. Gas giants are thought to consist of an outer layer of molecular hydrogen, surrounding a layer of liquid metallic hydrogen, with a probable molten core with a rocky composition.

Jupiter and Saturn's outermost portion of the hydrogen atmosphere has many layers of visible clouds that are mostly composed of water and ammonia. The layer of metallic hydrogen makes up the bulk of each planet, and is referred to as "metallic" because the very high pressure turns hydrogen into an electrical conductor. The core is thought to consist of heavier elements at such high temperatures (20,000 K) and pressures that their properties are poorly understood.

Ice Giants

Ice giants have distinctly different interior compositions from gas giants. The Solar System's ice giants, Uranus and Neptune, have a hydrogen-rich atmosphere that extends from the cloud tops down to about 80% (Uranus) or 85% (Neptune) of their radius. Below this, they are predominantly "icy", i.e. consist mostly of water, methane, and ammonia. There is also some rock and gas, but various proportions of ice–rock–gas could mimic pure ice, so that the exact proportions are unknown.

Uranus and Neptune have very hazy atmospheric layers with small amounts of methane, giving them aquamarine colors; light blue and ultramarine respectively. Both have magnetic fields that are sharply inclined to their axes of rotation.

Unlike the other giant planets, Uranus has an extreme tilt that causes its seasons to be severely pronounced. The two planets also have other subtle but important differences. Uranus has more hydrogen and helium than Neptune despite being less massive overall. Neptune is therefore denser and has much more internal heat and a more active atmosphere. The Nice model, in fact, suggests that Neptune formed closer to the Sun than Uranus did, and should therefore have more heavy elements.

Massive Solid Planets

Massive solid planets can also exist. For example, Kepler-10c is composed of mostly of rock with possibly up to 20% high-pressure water ice, but does have a hydrogen-dominated envelope.

Solid planets up to thousands of Earth masses may be able to form around massive stars (B-type and O-type stars; 5–120 solar masses), where the protoplanetary disk would contain enough heavy elements. Also, these stars have high UV radiation and winds that could photoevaporate the gas in the disk, leaving just the heavy elements. For comparison, Neptune's mass equals 17 Earth masses, Jupiter has 318 Earth masses, and the 13 Jupiter-mass limit used in the IAU's working definition of an exoplanet equals approximately 4000 Earth masses.

Extrasolar Giant Planets

An artist's conception of 79 Ceti b, the first extrasolar giant planet found with a minimum mass less than Saturn

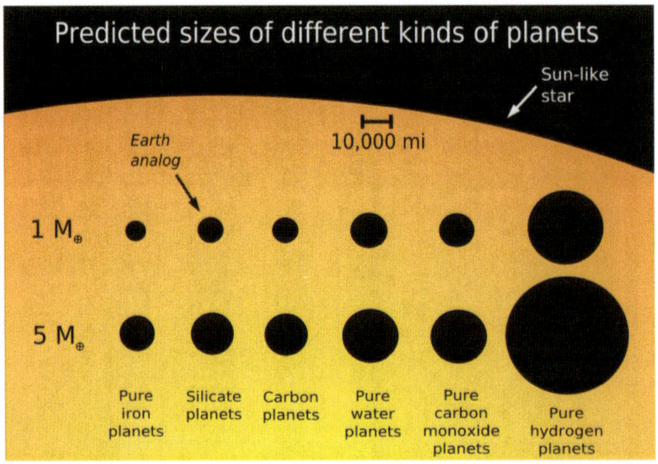

Comparison of sizes of planets of a given mass with different compositions

Because of the limited techniques currently available to detect exoplanets, many of those found to date have been of a size associated, in the Solar System, with giant planets. Because these large planets are inferred to share more in common with Jupiter than with the other giant planets, some have claimed that "jovian planet" is a more accurate term for them. Many of the exoplanets are much closer to their parent stars and hence much hotter than the giant planets in the Solar System, making it possible that some of those planets are a type not observed in the Solar System. Considering the relative abundances of the elements in the universe (approximately 98% hydrogen and helium) it would be surprising to find a predominantly rocky planet more massive than Jupiter. On the other hand, models of planetary-system formation have suggested that giant planets would be inhibited from forming as close to their stars as many of the extrasolar giant planets have been observed to orbit.

Atmospheres

The bands seen in the atmosphere of Jupiter are due to counter-circulating streams of material called zones and belts, encircling the planet parallel to its equator. The zones are the lighter bands, and are at higher altitudes in the atmosphere. They have an internal updraft and are high-pressure regions. The belts are the darker bands, are lower in the atmosphere, and have an internal downdraft. They are low-pressure regions. These structures are somewhat analogous to the high and low-pressure cells in Earth's atmosphere, but they have a very different structure latitudinal bands that circle the entire planet, as opposed to small confined cells of pressure. This appears to be a result of the rapid rotation and underlying symmetry of the planet. There are no oceans or landmasses to cause local heating and the rotation speed is much higher than that of Earth.

There are smaller structures as well: spots of different sizes and colors. On Jupiter, the most noticeable of these features is the Great Red Spot, which has been present for at least 300 years. These structures are huge storms. Some such spots are thunderheads as well.

Exoplanet

An extrasolar planet, also called an exoplanet, is a planet that orbits a star (i.e. is part of a solar system) other than our own. Our Solar System is only one among billions and many of them most likely have their own system of planets. As early as the sixteenth century, there have been astronomers who hypothesized of the existence of extrasolar planets.

The first recorded mention was made by Italian philosopher Giordano Bruno, an early supporter of the Copernican theory. In addition to supporting the idea that the Earth and other planets orbit the Sun (heliocentrism), he put forward the view that the fixed stars are similar to the Sun and are likewise accompanied by planets.

In the eighteenth century, Isaac Newton made a similar suggestion in the "General Scholium" section which concludes his Principia. Making a comparison to the Sun's planets, he wrote "And if the fixed stars are the centres of similar systems, they will all be constructed according to a similar design and subject to the dominion of One."

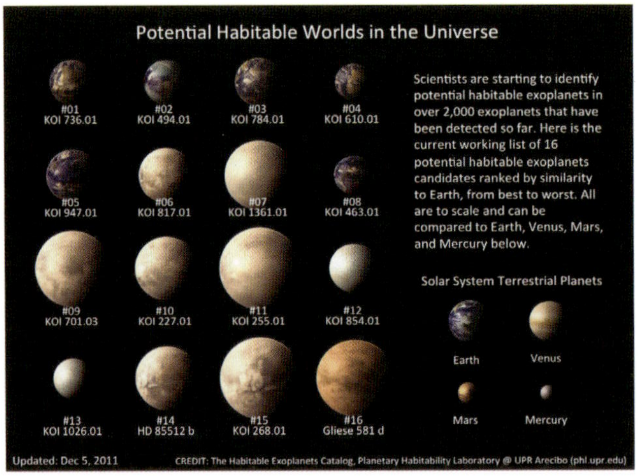

Since Newton's time, various claims have been made, but which were rejected by the scientific community as false positives. In the 1980's, some astronomers claimed that they had identified a some extrasolar planets in nearby star systems, but were unable to confirm their existence until years later.

First Discoveries

One of the reasons why extrasolar planets are so difficult to detect is because they are even fainter than the stars they orbit. Additionally, these stars give off light that "washes" the planets out – i.e. obscures them from direct observation. As a result, the first discovery was not made until 1992 when astronomers Aleksander Wolszczan and Dale Frail – using the Arecibo Observatory in Puerto Rico – observed several terrestrial-mass planets orbiting the pulsar PSR B1257+12.

It was not until 1995 that the first confirmation of an exoplanet orbiting a main-sequence star was made. In this case, the planet observed was 51 Pegasi b, a giant planet found in a four-day orbit around the Sun-like star 51 Pegasi (approx 51 light years from our Sun).

Initially, most of the planets detected were gas giants similar to, or larger than, Jupiter – which led to the term "Super-Jupiter" being coined. Far from suggesting that gas giants were more common than rocky (i.e. "Earth-like") planets, these findings were simply due to the fact that Jupiter-sized planets are simply easier to detect because of their size.

The Kepler Mission

Named after the Renaissance astronomer Johannes Kepler, the Kepler space observatory was launched by NASA on March 7th, 2009 for the purpose of discovering Earth-like planets orbiting other stars. As part of NASA's Discovery Program, a series of relatively low-cost project focused on scientific research, Kepler's mission is to survey a portion of our region of the Milky Way to find evidence of extrasolar planets and estimate how many stars in our galaxy have planetary systems.

Relying on the Transit Method of detection (see below), Kepler's sole instrument is a photometer that continually monitors the brightness of over 145,000 main sequence stars in a fixed field of view. This data is transmitted back to Earth where it is analyzed by scientists to look for any signs of periodic dimming caused by extrasolar planets transiting (passing) in front of their host star.

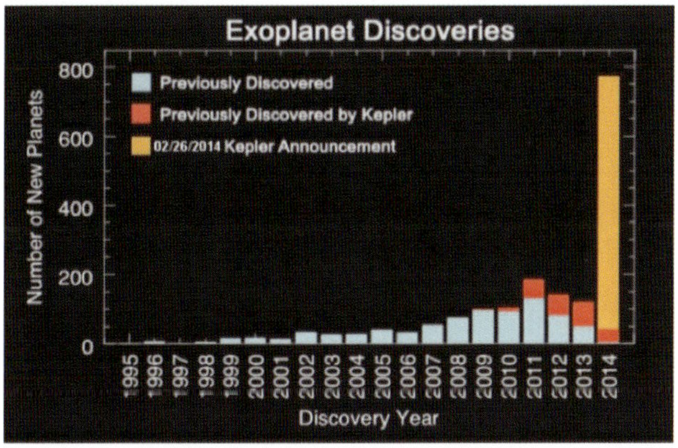

Histogram showing the number of planet discoveries by year for roughly the past two decades, broken down based on detection method

As of January 2015, Kepler and its follow-up observations had found 1,013 confirmed exoplanets in about 440 stellar systems, along with a further 3,199 unconfirmed planet candidates. In November 2013, astronomers reported, based on Kepler space mission data, that there could be as many as 40 billion Earth-sized planets orbiting in the habitable zones of Sun-like and red dwarf stars within the Milky Way. It is estimated that 11 billion of these planets may be orbiting Sun-like stars.

The initial planned lifetime of the Kepler mission was 3.5 years, but greater-than-expected results led to the mission being extended. In 2012, the mission was expected to last until 2016, but this changed due to the failure of one the spacecraft's reaction wheels – which are used for pointing the spacecraft. On May 11, 2013, a second of four reaction wheels failed, disabling the collection of science data and threatening the continuation of the mission.

On August 15, 2013, NASA announced that they had given up trying to fix the two failed reaction wheels and modified the mission accordingly. Rather than scrap Kepler, NASA proposed changing the mission to utilizing Kepler to detect habitable planets around smaller, dimmer red dwarf stars. This proposal, which became known as K2 "Second Light", was approved on May 16th, 2014.

Habitable Planets

Diagram showing the habitable zone of the Solar System (upper row) and in the Gliese 581 system (lower row)

The discovery of exoplanets has also intensified interest in the search for extraterrestrial life, particularly for those that orbit in the host star's habitable zone. Also known as the "goldilocks zone," this is the region of the solar system where conditions are warm enough (but not too warm) so that it is possible for liquid water (and therefore life) to exist on the planet's surface.

The first planet confirmed by Kepler to have an average orbital distance that placed it within its star's habitable zone was Kepler-22b. This planet is located about 600 light years from Earth in the constellation of Cygnus, and was first observed on May 12th, 2009, and then confirmed on Dec 5th, 2011. Based on all the data obtained, scientists believe that this world is roughly 2.4 times the radius of Earth, and is likely covered in oceans or has a liquid or gaseous outer shell.

Prior to the deployment of Kepler, the vast majority of confirmed exoplanets fell into the category of Jupiter-sized or larger. However, as of March 2014, Kepler has identified more than 2,900 planetary candidates, many of them falling into the categories of Earth-size or "Super-Earth" size. Many of these are located in the habitable zone of their parent stars, and some even around Sun-like stars.

Methodology

The first exoplanet was detected on 6 October 1995, and was named 51 Pegasi b. When an extrasolar planet is observed to transit its parent star, astronomers are able to assess some physical properties of the planet from an interstellar distance, including planetary mass and size, which in turn provide fundamental constraints on models of their physical structure. Furthermore, such events afford the opportunity to study the dynamics and chemistry of its atmosphere.

Statistical surveys and individual characterization are the keys to addressing the fundamental questions in exoplanetology. As of August 2016, varying techniques have been used to discover 3,502 exoplanets. Documenting the properties of a large sample exoplanets at various ages, orbiting their parent stars of various types, will contribute to increased understanding or better models of planetary formation (accretion), geological evolution, orbit migration, and their potential habitability. Characterizing the atmospheres of extrasolar planets is the new frontier in exoplanetary science.

Detection Techniques

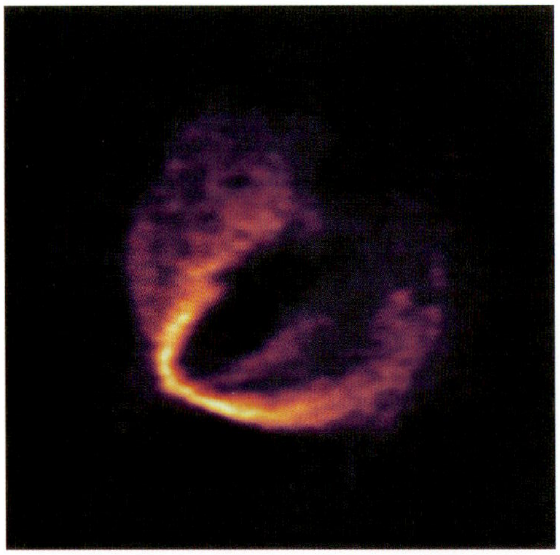

Measuring the flow of gas within a protoplanetary disc allows the detection of exoplanets

About 97% of all the confirmed exoplanets have been discovered by indirect techniques of detection, mainly by radial velocity measurements and transit monitoring techniques. The following methods have proved successful for discovering a new planet or confirming an already discovered planet:

- Radial velocity
- Gravitational microlensing
- Direct imaging
- Polarimetry
- Astrometry
- Transit photometry
    - Reflection/emission modulations
    - Light variations due to relativistic beaming
    - Light variations due to ellipsoidal variations
- Timing variations
    - Pulsar timing
    - Variable star timing
    - Transit timing variation method
    - Transit duration variation method
    - Eclipsing binary minima timing

Formation and Evolution

Planets form within a few tens of millions of years of their star forming. The planets of the Solar System can only be observed in their current state, but observations of different planetary systems of varying ages allows us to observe planets at different stages of evolution. Available observations range from young proto-planetary disks where planets are still forming to planetary systems of over 10 Gyr old. When terrestrial planets form in a gaseous protoplanetary disk, they have hydrogen envelopes that cool and contract over time and, depending on the mass of the planet, some or all of the hydrogen is eventually lost to space. This means that even terrestrial planets may start off with large radii if they form early enough. An example is Kepler-51b which has only about twice the mass of Earth but is almost the size of Saturn which is a hundred times the mass of Earth. Kepler-51b is quite young at a few hundred million years old.

Planet-hosting Stars

There is at least one planet on average per star. About 1 in 5 Sun-like stars have an "Earth-sized" planet in the habitable zone.

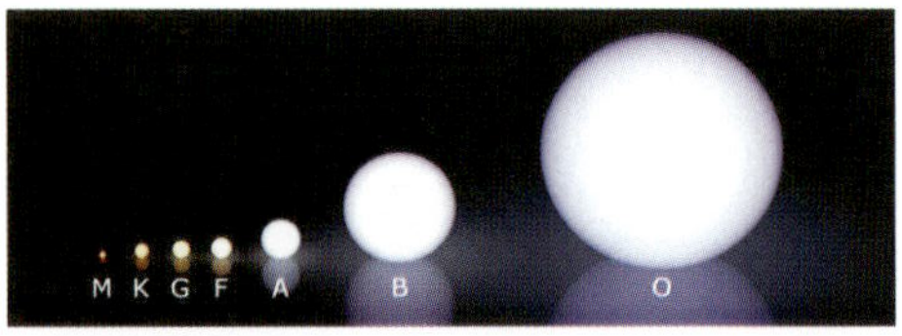

The Morgan-Keenan spectral classification

Artist's impression of exoplanet orbiting two stars

Most known exoplanets orbit stars roughly similar to the Sun, i.e. main-sequence stars of spectral categories F, G, or K. Lower-mass stars (red dwarfs, of spectral category M) are less likely to have planets massive enough to be detected by the radial-velocity method. Despite this, several tens of planets around red dwarfs have been discovered by the *Kepler* spacecraft, which uses the transit method to detect smaller planets.

Using data from Kepler, a correlation has been found between the metallicity of a star and the probability that the star host planets. Stars with higher metallicity are more likely to have planets, especially giant planets, than stars with lower metallicity.

Some planets orbit one member of a binary star system, and several circumbinary planets have been discovered which orbit around both members of binary star. A few planets in triple star systems are known and one in the quadruple system Kepler-64.

General Features

Color and Brightness

In 2013 the color of an exoplanet was determined for the first time. The best-fit albedo measurements of HD 189733b suggest that it is deep dark blue. Later that same year, the colors of several other exoplanets were determined, including GJ 504 b which visually has a magenta color, and Kappa Andromedae b, which if seen up close would appear reddish in color.

The apparent brightness (apparent magnitude) of a planet depends on how far away the observer is, how reflective the planet is (albedo), and how much light the planet receives from its star, which depends on how far the planet is from the star and how bright the star is. So, a planet with a low albedo that is close to its star can appear brighter than a planet with high albedo that is far from the star.

The darkest known planet in terms of geometric albedo is TrES-2b, a hot Jupiter that reflects less than 1% of the light from its star, making it less reflective than coal or black acrylic paint. Hot

Jupiters are expected to be quite dark due to sodium and potassium in their atmospheres but it is not known why TrES-2b is so dark it could be due to an unknown chemical compound.

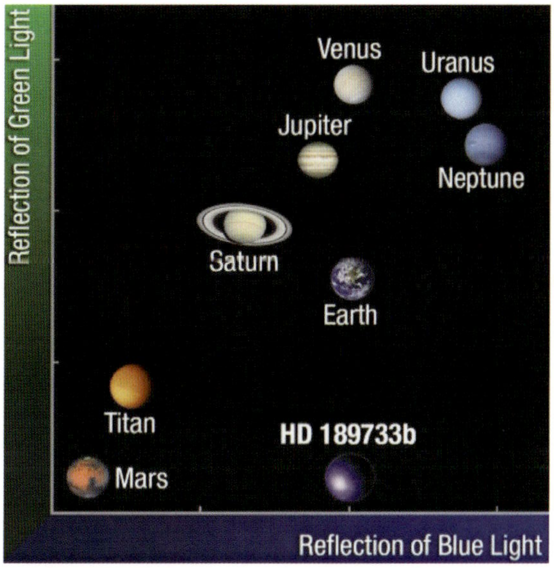

This color–color diagram compares the colors of planets in the Solar System to exoplanet HD 189733b. The exoplanet's deep blue color is produced by silicate droplets, which scatter blue light in its atmosphere

For gas giants, geometric albedo generally decreases with increasing metallicity or atmospheric temperature unless there are clouds to modify this effect. Increased cloud-column depth increases the albedo at optical wavelengths, but decreases it at some infrared wavelengths. Optical albedo increases with age, because older planets have higher cloud-column depths. Optical albedo decreases with increasing mass, because higher-mass giant planets have higher surface gravities, which produces lower cloud-column depths. Also, elliptical orbits can cause major fluctuations in atmospheric composition, which can have a significant effect.

There is more thermal emission than reflection at some near-infrared wavelengths for massive and/or young gas giants. So, although optical brightness is fully phase-dependent, this is not always the case in the near infrared.

Temperatures of gas giants reduce over time and with distance from their star. Lowering the temperature increases optical albedo even without clouds. At a sufficiently low temperature, water clouds form, which further increase optical albedo. At even lower temperatures ammonia clouds form, resulting in the highest albedos at most optical and near-infrared wavelengths.

Magnetic Field

In 2014, a magnetic field around HD 209458 b was inferred from the way hydrogen was evaporating from the planet. It is the first (indirect) detection of a magnetic field on an exoplanet. The magnetic field is estimated to be about one tenth as strong as Jupiter's.

Interaction between a close-in planet's magnetic field and a star can produce spots on the star in a similar way to how the Galilean moons produce aurorae on Jupiter. Auroral radio emissions could be detected with radio telescopes such as LOFAR. The radio emissions could enable determination of the rotation rate of a planet which is difficult to detect otherwise.

Earth's magnetic field results from its flowing liquid metallic core, but in massive super-Earths with high pressure, different compounds may form which do not match those created under terrestrial conditions. Compounds may form with greater viscosities and high melting temperatures which could prevent the interiors from separating into different layers and so result in undifferentiated coreless mantles. Forms of magnesium oxide such as $MgSi_3O_{12}$ could be a liquid metal at the pressures and temperatures found in super-Earths and could generate a magnetic field in the mantles of super-Earths.

Hot Jupiters have been observed to have a larger radius than expected. This could be caused by the interaction between the stellar wind and the planet's magnetosphere creating an electric current through the planet that heats it up causing it to expand. The more magnetically active a star is the greater the stellar wind and the larger the electric current leading to more heating and expansion of the planet. This theory matches the observation that stellar activity is correlated with inflated planetary radii.

Plate Tectonics

In 2007, two independent teams of researchers came to opposing conclusions about the likelihood of plate tectonics on larger super-Earths with one team saying that plate tectonics would be episodic or stagnant and the other team saying that plate tectonics is very likely on super-Earths even if the planet is dry.

If super-Earths have more than 80 times as much water as Earth then they become ocean planets with all land completely submerged. However, if there is less water than this limit, then the deep water cycle will move enough water between the oceans and mantle to allow continents to exist.

Volcanism

Large surface temperature variations on 55 Cancri e have been attributed to possible volcanic activity releasing large clouds of dust which blanket the planet and block thermal emissions.

Rings

The star 1SWASP J140747.93-394542.6 is orbited by an object that is circled by a ring system much larger than Saturn's rings. However, the mass of the object is not known; it could be a brown dwarf or low-mass star instead of a planet.

The brightness of optical images of Fomalhaut b could be due to starlight reflecting off a circumplanetary ring system with a radius between 20 and 40 times that of Jupiter's radius, about the size of the orbits of the Galilean moons.

The rings of the Solar System's gas giants are aligned with their planet's equator. However, for exoplanets that orbit close to their star, tidal forces from the star would lead to the outermost rings of a planet being aligned with the planet's orbital plane around the star. A planet's innermost rings would still be aligned with the planet's equator so that if the planet has a tilted rotational axis, then the different alignments between the inner and outer rings would create a warped ring system.

Moons

In December 2013 a candidate exomoon of a rogue planet was announced. No exomoons have been confirmed so far.

Atmospheres

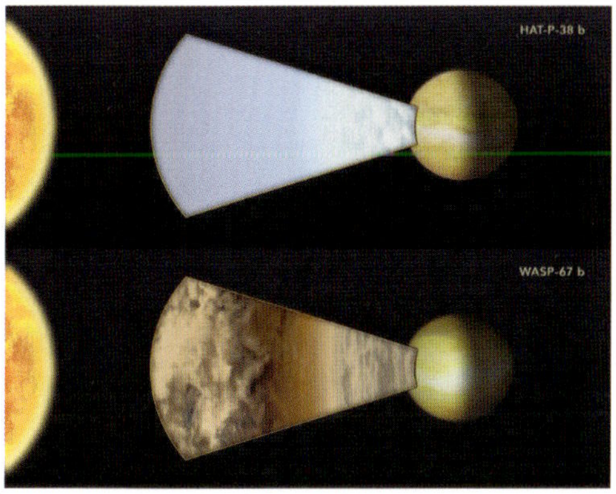

Clear versus cloudy atmospheres on two exoplanets

Atmospheres have been detected around several exoplanets. The first to be observed was HD 209458 b in 2001.

KIC 12557548 b is a small rocky planet, very close to its star, that is evaporating and leaving a trailing tail of cloud and dust like a comet. The dust could be ash erupting from volcanos and escaping due to the small planet's low surface-gravity, or it could be from metals that are vaporized by the high temperatures of being so close to the star with the metal vapor then condensing into dust.

In June 2015, scientists reported that the atmosphere of GJ 436 b was evaporating, resulting in a giant cloud around the planet and, due to radiation from the host star, a long trailing tail 14×10^6 km (9×10^6 mi) long.

In May 2017, glints of light from Earth, seen as twinkling from an orbiting satellite a million miles away, were found to be reflected light from ice crystals in the atmosphere. The technology used to determine this may be useful in studying the atmospheres of distant worlds, including those of exoplanets.

Insolation Pattern

Tidally locked planets in a 1:1 spin–orbit resonance would have their star always shining directly overhead on one spot which would be hot with the opposite hemisphere receiving no light and being freezing cold. Such a planet could resemble an eyeball with the hotspot being the pupil. Planets with an eccentric orbit could be locked in other resonances. 3:2 and 5:2 resonances would result in a double-eyeball pattern with hotspots in both eastern and western hemispheres. Planets with both an eccentric orbit and a tilted axis of rotation would have more complicated insolation patterns.

As more planets are discovered, the field of exoplanetology continues to grow into a deeper study of extrasolar worlds, and will ultimately tackle the prospect of life on planets beyond the Solar System. At cosmic distances, life can only be detected if it is developed at a planetary scale and strongly modified the planetary environment, in such a way that the modifications cannot be explained by classical physico-chemical processes (out of equilibrium processes). For example, molecular oxygen (O_2) in the atmosphere of Earth is a result of photosynthesis by living plants and many kinds of microorganisms, so it can be used as an indication of life on exoplanets, although small amounts of oxygen could also be produced by non-biological means. Furthermore, a potentially habitable planet must orbit a stable star at a distance within which planetary-mass objects with sufficient atmospheric pressure can support liquid water at their surfaces.

Dwarf Planet

The term dwarf planet has been tossed around a lot in recent years. As part of a three-way categorization of bodies orbiting the Sun, the term was adopted in 2006 due to the discovery of objects beyond the orbit of Neptune that were comparable in size to Pluto. Since then, it has come to be used to describe many objects in our Solar System, upending the old classification system that claimed there were nine planets.

The term has also led to its fair share of confusion and controversy, with many questioning its accuracy and applicability to bodies like Pluto. Nevertheless, the IAU currently recognizes five bodies within our Solar System as dwarf planets, six more could be recognized in the coming years, and as many as 200 or more could exist within the expanse of the Kuiper Belt.

According to the definition adopted by the IAU in 2006, a dwarf planet is, "a celestial body orbiting a star that is massive enough to be rounded by its own gravity but has not cleared its neighbouring region of planetesimals and is not a satellite. More explicitly, it has to have sufficient mass to overcome its compressive strength and achieve hydrostatic equilibrium."

In essence, the term is meant to designate any planetary-mass object that is neither a planet nor a natural satellite that fits two basic criteria. For one, it must be in direct orbit of the Sun and not be a moon around another body. Second, it must be massive enough for it to have become spherical in shape under its own gravity. And, unlike a planet, it must have not cleared the neighbourhood around its orbit.

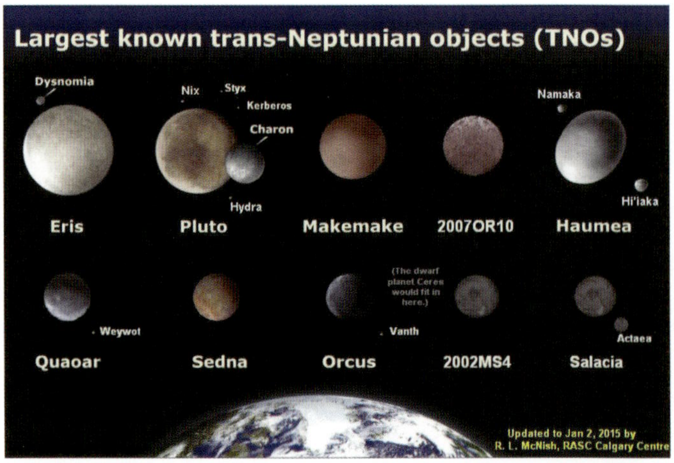

The largest known trans-Neptunian objects (TNO), shown to scale

Size and Mass

In order for a body to be become rounded, it must be sufficiently massive, to the point that its own gravity is the dominant force effecting it. Here, the internal pressure created by this mass would cause a surface to achieve plasticity, allowing high elevations to sink and hollows to fill in. This does not occur with smaller bodies that are less than a few km in diameter (such as asteroids), which are dominated by forces outside of their own gravity forces and tend to maintain irregular shapes.

Meanwhile, bodies that measure a few kilometers across – where their gravity is more significant but not dominant – tend to be spheroid or "potato-shaped". The bigger the body is, the higher its internal pressure, until the pressure is sufficient to overcome its internal compressive strength and it achieves hydrostatic equilibrium. At this point, a body is as round as it can possibly be, given its rotation and tidal effects. This is the defining limit of a dwarf planet.

However, rotation can also affect the shape of a dwarf planet. If the body does not rotate, it will be a sphere. But the faster it does rotate, the more oblate or even scalene it becomes. The extreme example of this is Haumea, which is twice as long along its major axis as it is at the poles. Tidal forces also cause a body's rotation to gradually become tidally locked, such that it always presents the same face to its companion. An extreme example of this is the Pluto-Charon system, where both bodies are tidally locked to each other.

The upper and lower size and mass limits of dwarf planets have not been specified by the IAU. And while the lower limit is defined as the achievement of a hydrostatic equilibrium shape, the size or mass at which an object attains this shape depends on its composition and thermal history.

For example, bodies made of rigid silicates (such as rocky asteroids) should achieve hydrostatic equilibrium at a diameter of approx. 600 km and a mass of 3.4×10^{20} kg. For a body made of less rigid water ice, the limit would closer to 320 km and 10^{19} kg. As a result, no specific standard currently exists for defining a dwarf planet based on either its size or mass, but is instead more generally defined based on its shape.

Orbital Dominance

In addition to hydrostatic equilibrium, many astronomers have insisted that a distinction between planets and dwarf planets be made based on the inability of the latter to "clear the neighbourhood around their orbits". In short, planets are able to remove smaller bodies near their orbits by collision, capture, or gravitational disturbance (or establish orbital resonances that prevent collisions), whereas dwarf planets do not have the requisite mass to do this.

To calculate the likelihood of a planet clearing its orbit, planetary scientists Alan Stern and Harold F. Levison (the former of whom is the principal investigator of the New Horizons mission to Pluto and the Chief Scientist at Moon Express) introduced a parameter they designated as ? (lambda).

This parameter expresses the likelihood of an encounter resulting in a given deflection of an object's orbit. The value of this parameter in Stern's model is proportional to the square of the mass and inversely proportional to the period, and can be used to estimate the capacity of a body to clear the neighbourhood of its orbit.

Astronomers like Steven Soter, the scientist-in-residence for NYU and a Research Associate at the American Museum of Natural History, have advocated using this parameter to differentiate between planets and dwarf planets. Soter has also proposed a parameter he refers to as the *planetary discriminant* – designated as μ (mu) – which is calculated by dividing the mass of the body by the total mass of the other objects that share its orbit.

Recognized and Possible Dwarf Planets

There are currently five dwarf planets: Pluto, Eris, Makemake, Haumea, and Ceres. Only Ceres and Pluto have been observed enough to indisputably fit into the category. The IAU decided that unnamed Trans-Neptunian Objects (TNOs) with an absolute magnitude brighter than +1 (and a mathematically delimited minimum diameter of 838 km) are to be named as dwarf planets.

Possible candidates that are currently under consideration include Orcus, 2002 MS_4, Salacia, Quaoar, 2007 OR_{10}, and Sedna. All of these objects are located in the Kuiper Belt or the Scattered Disc; with the exception of Sedna, which is a detached object – a special class that applies to dynamic TNOs in the outer Solar System.

It is possible that there are another 40 known objects in the solar system that could be rightly classified as dwarf planets. Estimates are that up to 200 dwarf planets may be found when the entire region known as the Kuiper belt is explored, and that the number may exceed 10,000 when objects scattered outside the Kuiper belt are considered.

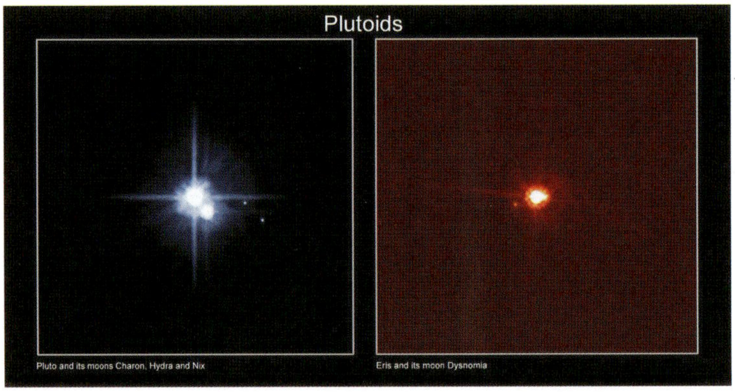

Pluto and moons Charon, Hydra and Nix (left)
compared to the dwarf planet Eris and its moon Dysmonia (right)

Ceres

Ceres is the earliest known and smallest of the current category of dwarf planets. Sicilian astronomer Giuseppe Piazzi discovered Ceres in 1801 based on the prediction that the gap between Mars and Jupiter contained a missing planet. It is only 590 miles (950 km) in diameter and has a mass of just 0.015 percent that of Earth.

In fact, Ceres is so small that it is classified as both a dwarf planet and an asteroid, and is often named in scientific literature as one of the largest asteroids in the solar system. Although it makes up approximately a fourth of the mass of the asteroid belt, it is still 14 less massive than Pluto.

Unlike its asteroid neighbours, Ceres has a nearly round body. The rocky dwarf planet may have water ice beneath its crust. In 2014, the European Space Agency's Herschel Space Observatory detected water vapor spewing from two regions on Ceres.

NASA's robotic Dawn mission arrived at Ceres in 2015. The mission has shown many interesting features on its surface, ranging from various bright spots to a 4-mile-high (6.5-kilometer-high) mountain. (Another mission, the European Space Agency's Herschel Space Observatory, spotted evidence of water vapor in 2014.)

Pluto

Pluto is the most well known of the dwarf planets. Since its discovery in 1930 and until 2006, it had been classified as the ninth planet from the sun. Pluto's orbit was so erratic, however, that at times it was closer to the sun than the eighth planet, Neptune. In 2006, with the discovery of several other rocky bodies similar in size or larger than Pluto, the IAU decided to re-classify Pluto as a dwarf planet.

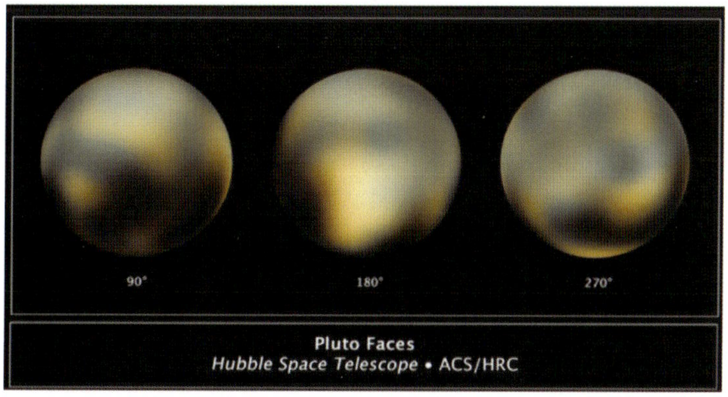

This is the most detailed view to date of the entire surface of the dwarf planet Pluto, as constructed from multiple NASA Hubble Space Telescope photographs taken from 2002 to 2003

Despite its small size 0.2 percent the mass of Earth and only 10 percent the mass of Earth's moon Pluto's gravity is enough to capture five moons of its own. The pairing between Pluto and its largest moon, Charon, is known as a binary system, because both objects are orbiting around a central point that is not within the mass of Pluto.

NASA's New Horizons mission flew by Pluto in 2015 and revealed a wealth of surprises. This included zones that are bereft of craters (indicating the surface is relatively young), mountains that are likely as high as 11,000 feet (3,500 meters), and even haze above the dwarf planet's surface.

Eris

When it was first discovered, Eris was thought to be the largest of the dwarf planets, with a mass 27 percent larger than that of Pluto and a diameter of approximately 1,400 to 1,500 miles (2,300 to 2,400 km). It was the discovery of Eris that prompted the IAU to reconsider the definition of a planet. Further observation went on to suggest that the dwarf planet is slightly smaller than Pluto.

The orbit of Eris is very erratic, crossing that of Pluto and nearly intersecting the orbit of Neptune, but is still more than three times larger than Pluto's orbit. It takes 557 years for Eris to orbit the sun. At its farthest point from the sun, a point that is also called its aphelion, Eris and its satellite

Dysmonia travel far beyond the Kuiper Belt. The surface of Eris is likely nitrogen and methane-rich, but in a thin (1 millimeter) layer across the surface. Some scientists suggest the surface is the condensed atmosphere of Eris, which expands into gas when the dwarf planet is closer to the sun.

An early artist's interpretation of the dwarf planet Makemake beyond Pluto

Haumea and Makemake

Haumea and Makemake are the most recently named dwarf planets in the solar system.

Haumea is unique because of its ellipsoid shape, only just meeting the hydrostatic equilibrium criteria for dwarf planet status. The elongated shape of the dwarf planet is due to its rapid rotational spin, not a lack of mass, which is about one-third that of Pluto. The cigar-shaped dwarf planet rotates on its axis every four hours, likely a result of a collision. The odd object also hosts a red spot and a layer of crystalline ice. Finally, Haumea is the only object in the Kuiper belt other than Pluto known to host more than one moon.

A moon was discovered around Makemake in 2016, more than a decade after the dwarf planet itself was found. Its diameter is known to be about two-thirds that of Pluto, and the newly found moon will allow for measurements of its mass. Makemake is also of value to the astronomical community, as it is another reason for the reconsideration of the definition of a planet. Its comparable mass and diameter to Pluto would grant it planet status if Pluto wasn't also stripped of that title.

Dwarf Planets as 'Plutoids'

Pluto, Eris, Haumea and Makemake are all known as "plutoids," unlike the asteroidal dwarf planetoid Ceres. A plutoid is a dwarf planet with an orbit outside that of Neptune. Plutoids are sometimes also referred to as "ice dwarfs" due to their diminutive size and cold surface temperatures.

The outer planets show evidence of interaction with plutoids. Triton, the largest moon of Neptune, is likely a captured plutoid, and it is even possible that the odd tilt of Uranus on its axis is due to a collision with a plutoid. Similarly to dwarf planets, there are potentially hundreds of plutoid objects in the solar system that have yet to be given official status.

References

- Hermann Hunger, ed. (1992). Astrological reports to Assyrian kings. State Archives of Assyria. 8. Helsinki University Press. ISBN 951-570-130-9
- Lyttleton, Raymond A. (1936). "On the possible results of an encounter of Pluto with the Neptunian system". Monthly Notices of the Royal Astronomical Society. 97 (2): 108–115. Bibcode:1936MNRAS..97..108L. doi:10.1093/mnras/97.2.108
- Terrestrial-planet-50289: universetoday.com, Retrieved 19 June 2018
- Stern, S. Alan (2004-03-22). "Gravity Rules: The Nature and Meaning of Planethood". SpaceDaily. Retrieved 2008-08-23
- Saumon, D.; Hubbard, W. B.; Burrows, A.; Guillot, T.; et al. (1996). "A Theory of Extrasolar Giant Planets". Astrophysical Journal. 460: 993–1018. arXiv:astro-ph/9510046. Bibcode:1996ApJ...460..993S. doi:10.1086/177027
- What-are-extrasolar-planets-1706656300: gizmodo.com, Retrieved 29 June 2018
- Evans, James (1998). The History and Practice of Ancient Astronomy. Oxford University Press. pp. 296–7. ISBN 978-0-19-509539-5. Retrieved 2008-02-04
- Whipple, Fred (1964). "The History of the Solar System". Proceedings of the National Academy of Sciences of the United States of America. 52 (2): 565–594. Bibcode:1964PNAS...52..565W. doi:10.1073/pnas.52.2.565. PMC 300311. PMID 16591209
- What-is-a-dwarf-planet-72717: universetoday.com, Retrieved 31 March 2018
- Green, D. W. E. (2006-09-13). "(134340) Pluto, (136199) Eris, and (136199) Eris I (Dysnomia)" (PDF). Central Bureau for Astronomical Telegrams, International Astronomical Union. Circular No. 8747. Archived from the original on June 24, 2008. Retrieved 2011-07-05
- Croswell, K. (1997). Planet Quest: The Epic Discovery of Alien Solar Systems. The Free Press. p. 57. ISBN 978-0-684-83252-4
- Dwarf-planets-facts-solar-system-sdcmp-15216: space.com, Retrieved 21 May 2018

7

Understanding Galaxies

A galaxy is a system of stars and stellar remnants, dust, interstellar gas and dark matter. Some of the issues addressed in astrophysics regarding the study of galaxies are galaxy formation and evolution, interacting galaxy, galaxy classification, magellanic clouds, etc. which have been covered in this chapter.

A galaxy is a massive, gravitationally bound system consisting of stars, an interstellar medium of gas, plasma, and dust, and dark matter. Typical galaxies range from dwarfs with as few as ten million (10^7) stars up to giants with one trillion (10^{12}) stars, all orbiting a common centre of mass. Galaxies can also contain many multiple star systems, star clusters, and various interstellar clouds. The Solar System is located in the Milky Way galaxy.

Galaxies have been categorized according to their apparent shapes. They include elliptical, spiral, and peculiar galaxies (with irregular or unusual shapes). Interactions between nearby galaxies, which may cause them to merge, may enhance star formation, producing a starburst galaxy.

The universe contains at least two trillion galaxies, ten times more than previously thought. Most galaxies are 1,000 to 100,000 parsecs in diameter and are usually separated by distances on the order of millions of parsecs (or megaparsecs). Intergalactic space is filled with a tenuous gas of average density less than one atom per cubic meter. Most galaxies are organized into clusters, which in turn can form larger groups called superclusters. The larger structures are generally arranged into sheets and filaments that surround immense voids in the universe.

Although not yet well understood, dark matter appears to account for around 90 percent of the mass of most galaxies. Observational data suggests that supermassive black holes may exist at the centre of many, if not all, galaxies. They are proposed to be the primary cause of active galactic nuclei found at the core of some galaxies. The Milky Way galaxy appears to harbor at least one such object within its nucleus.

The Milky Way

The Greek philosopher Democritus (450–370 B.C.E.) proposed that the bright band on the night sky known as the Milky Way might consist of distant stars. Aristotle (384-322 B.C.E.), however, believed the Milky Way to be caused by "the ignition of the fiery exhalation of some stars which were large, numerous and close together" and that the "ignition takes place in the upper part of the atmosphere, in the region of the world which is continuous with the heavenly motions." The Arabian astronomer, Alhazen (965-1037 C.E.), refuted this by making the first attempt at observing and measuring the Milky Way's parallax, and he thus "determined that because the Milky Way had no parallax, it was very remote from the earth and did not belong to the atmosphere."

The Persian astronomer, Abū Rayhān al-Bīrūnī (973-1048), proposed the Milky Way galaxy to be a collection of countless nebulous stars. Avempace (d. 1138) proposed the Milky Way to be made up of many stars but appears to be a continuous image due to the effect of refraction in the Earth's atmosphere. Ibn Qayyim Al-Jawziyya (1292-1350) proposed the Milky Way galaxy to be "a myriad of tiny stars packed together in the sphere of the fixed stars" and that these stars are larger than planets.

Actual proof of the Milky Way consisting of many stars came in 1610 when Galileo Galilei used a telescope to study the Milky Way and discovered that it is composed of a huge number of faint stars. In a treatise in 1755, Immanuel Kant, drawing on earlier work by Thomas Wright, speculated (correctly) that the Galaxy might be a rotating body of a huge number of stars held together by gravitational forces, akin to the solar system but on a much larger scale. The resulting disk of stars can be seen as a band on the sky from our perspective inside the disk.

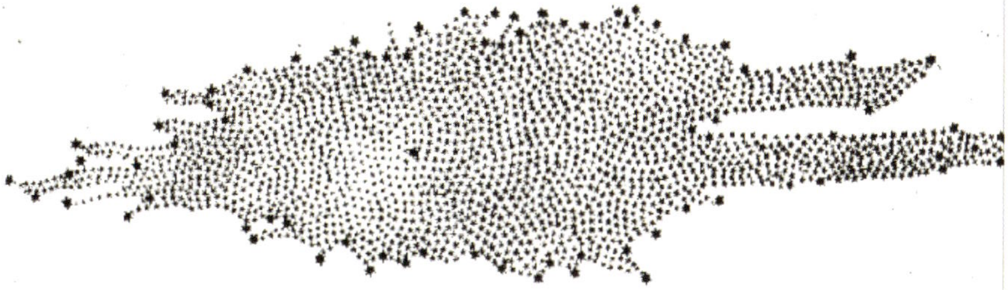

The shape of the Milky Way as deduced from star counts by William Herschel in 1785; the solar system was assumed to be near the centre

The first attempt to describe the shape of the Milky Way and the position of the Sun in it was carried out by William Herschel in 1785 by carefully counting the number of stars in different regions of the sky. He produced a diagram of the shape of the galaxy with the solar system close to the centre. Using a refined approach, Kapteyn in 1920 arrived at the picture of a small (diameter about 15 kiloparsecs) ellipsoid galaxy with the Sun close to the centre. Both analyses failed to take into account the absorption of light by interstellar dust present in the galactic plane, but after Robert Julius Trumpler quantified this effect in 1930 by studying open clusters, the present picture of our galaxy, the Milky Way, emerged.

Other Nebulae

Sketch of the Whirlpool Galaxy by Lord Rosse in 1845

In the tenth century, the Persian astronomer, Abd al-Rahman al-Sufi (known in the West as *Azophi*), made the earliest recorded observation of the Andromeda Galaxy, describing it as a "small cloud". Al-Sufi also identified the Large Magellanic Cloud, which is visible from Yemen, though not from Isfahan; it was not seen by Europeans until Magellan's voyage in the sixteenth century. These were the first galaxies other than the Milky Way to be observed from Earth. Al-Sufi published his findings in his *Book of Fixed Stars* in 964.

In 1054, the creation of the Crab Nebula resulting from the SN 1054 supernova was observed by Chinese and Arabian astronomers. The Crab Nebula itself was observed centuries later by John Bevis in 1731, followed by Charles Messier in 1758 and then by the Earl of Rosse in the 1840s.

Toward the end of the eighteenth century, Charles Messier compiled a catalog containing the 109 brightest nebulae (celestial objects with a nebulous appearance), later followed by a larger catalog of 5,000 nebulae assembled by William Herschel. In 1845, Lord Rosse constructed a new telescope and was able to distinguish between elliptical and spiral nebulae.

In 1917, Heber Curtis had observed the nova S Andromedae within the "Great Andromeda Nebula" (Messier object M31). Searching the photographic record, he found 11 more novae. Curtis noticed that these novae were, on average, 10 magnitudes fainter than those that occurred within our galaxy. As a result he was able to come up with a distance estimate of 150,000 parsecs. He became a proponent of the so-called "island universes" hypothesis, which holds that spiral nebulae are actually independent galaxies.

Photograph of the "Great Andromeda Nebula" from 1899, later identified as the Andromeda Galaxy

In 1920 the so-called Great Debate took place between Harlow Shapley and Heber Curtis, concerning the nature of the Milky Way, spiral nebulae, and the dimensions of the universe. To support his claim that the Great Andromeda Nebula was an external galaxy, Curtis noted the appearance of dark lanes resembling the dust clouds in the Milky Way, as well as the significant Doppler shift.

The matter was conclusively settled by Edwin Hubble in the early 1920s using a new telescope. He was able to resolve the outer parts of some spiral nebulae as collections of individual stars and identified some Cepheid variables, thus allowing him to estimate the distance to the nebulae: they were far too distant to be part of the Milky Way. In 1936 Hubble produced a classification system for galaxies that is used to this day, the Hubble sequence.

Modern Research

In 1944 Hendrik van de Hulst predicted microwave radiation at a wavelength of 21 cm resulting from interstellar atomic hydrogen gas; this radiation was observed in 1951. The radiation allowed for much improved study of the Milky Way Galaxy, since it is not affected by dust absorption and its Doppler shift can be used to map the motion of the gas in the Galaxy. These observations led to the postulation of a rotating bar structure in the centre of the Galaxy. With improved radio telescopes, hydrogen gas could also be traced in other galaxies.

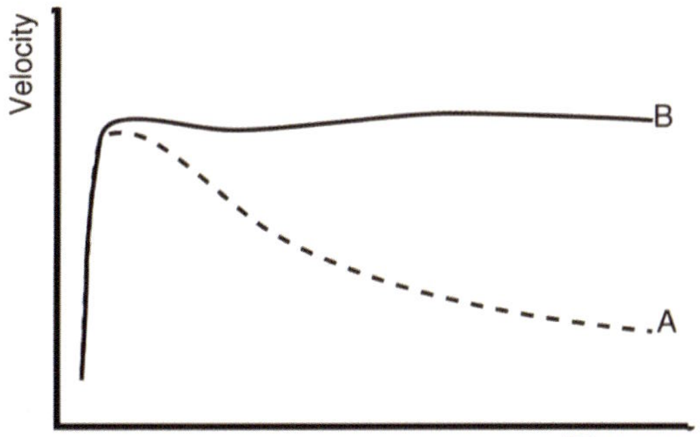

Rotation curve of a typical spiral galaxy: predicted (A) and observed (B). The distance is from the galactic core

In the 1970s it was discovered in Vera Rubin's study of the rotation speed of gas in galaxies that the total visible mass (from the stars and gas) does not properly account for the speed of the rotating gas. This galaxy rotation problem is thought to be explained by the presence of large quantities of unseen dark matter.

Beginning in the 1990s, the Hubble Space Telescope yielded improved observations. Among other things, it established that the missing dark matter in our galaxy cannot solely consist of inherently faint and small stars. In 1999, the Hubble Deep Field, an extremely long exposure of a relatively empty part of the sky, provided evidence that there are about 125 billion galaxies in the universe, and later with the new camera HST has observed 3,000 visible galaxies. Improved technology in detecting the spectra invisible to humans (radio telescopes, infrared cameras, and x-ray telescopes) allow detection of other galaxies that are not detected by Hubble. Particularly, galaxy surveys in the zone of avoidance (the region of the sky blocked by the Milky Way) have revealed a number of new galaxies.

Types and Morphology

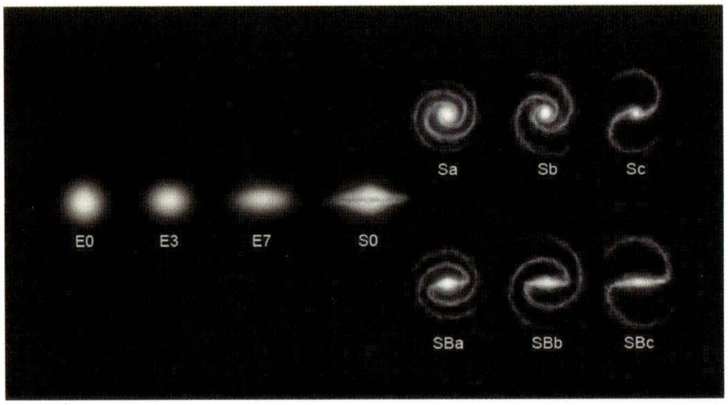

Types of galaxies according to the Hubble classification scheme. An *E* indicates a type of elliptical galaxy; an *S* is a spiral; and *SB* is a barred-spiral galaxy

Galaxies come in three main types: ellipticals, spirals, and irregulars. A slightly more extensive description of galaxy types based on their appearance is given by the Hubble sequence. Since the Hubble sequence is entirely based upon visual morphological type, it may miss certain important characteristics of galaxies such as star formation rate (in starburst galaxies) and activity in the core (in active galaxies).

Ellipticals

The giant elliptical galaxy ESO 325-G004

An elliptical galaxy is a type of galaxy having an approximately ellipsoidal shape and a smooth, nearly featureless image. Unlike flat spiral galaxies with organization and structure, elliptical galaxies are more three-dimensional, without much structure, and their stars are in somewhat random orbits around the centre.

Elliptical galaxies range in size from tens of millions to over one hundred trillion stars. Most elliptical galaxies are composed of older, low-mass stars, with a sparse interstellar medium and minimal star formation activity, and they tend to be surrounded by large numbers of globular clusters. Elliptical galaxies are believed to make up approximately 10–15 percent of galaxies in the Virgo Supercluster, and they are not the dominant type of galaxy in the universe overall. They are preferentially found close to the centres of galaxy clusters.

Originally, Edwin Hubble hypothesized that elliptical galaxies evolved into spiral galaxies. This theory was later discovered to be false, although the accretion of gas and smaller galaxies may build a disk around a pre-existing ellipsoidal structure. Stars found inside of elliptical galaxies are on average much older than stars found in spiral galaxies.

Spirals

The Sombrero Galaxy, an example of an unbarred spiral galaxy

Spiral galaxies consist of a rotating disk of stars and interstellar medium, along with a central bulge of generally older stars. Extending outward from the bulge are relatively bright arms. In the Hubble classification scheme, spiral galaxies are listed as type *S*, followed by a letter (*a*, *b*, or *c*) that indicates the degree of tightness of the spiral arms and the size of the central bulge. An *Sa* galaxy has tightly wound, poorly-defined arms and possesses a relatively large core region. At the other extreme, an *Sc* galaxy has open, well-defined arms and a small core region.

In spiral galaxies, the spiral arms have the shape of approximate logarithmic spirals, a pattern that can be theoretically shown to result from a disturbance in a uniformly rotating mass of stars. Like the stars, the spiral arms also rotate around the centre, but they do so with constant angular velocity. That means that stars pass in and out of spiral arms, with stars near the galactic core orbiting faster than the arms are moving while stars near the outer parts of the galaxy typically orbit more slowly than the arms. The spiral arms are thought to be areas of high density matter, or "density waves." As stars move through an arm, the space velocity of each stellar system is modified by the gravitational force of the higher density. (The velocity returns to normal after the stars depart on the other side of the arm.) This effect is akin to a "wave" of slowdowns moving along a highway full of moving cars. The arms are visible because the high density facilitates star formation, and therefore they harbor many bright and young stars.

NGC 1300, an example of a barred spiral galaxy

A majority of spiral galaxies have a linear, bar-shaped band of stars that extends outward to either side of the core, then merges into the spiral arm structure. In the Hubble classification scheme, these are designated by an *SB*, followed by a lower-case letter (*a*, *b* or *c*) that indicates the form of the spiral arms (in the same manner as the categorization of normal spiral galaxies). Bars are thought to be temporary structures that can occur as a result of a density wave radiating outward from the core, or else due to a tidal interaction with another galaxy. Many barred spiral galaxies are active, possibly as a result of gas being channeled into the core along the arms.

Our own galaxy is a large disk-shaped barred-spiral galaxy about 30 kiloparsecs in diameter and a kiloparsec in thickness. It contains about two hundred billion (2×10^{11}) stars and has a total mass of about six hundred billion (6×10^{11}) times the mass of the Sun.

Other Morphologies

Hoag's Object, an example of a ring galaxy

Peculiar galaxies are galactic formations that develop unusual properties due to tidal interactions with other galaxies. An example of this is the ring galaxy, which possesses a ring-like structure of stars and interstellar medium surrounding a bare core. A ring galaxy is thought to occur when a smaller galaxy passes through the core of a spiral galaxy. Such an event may have affected the Andromeda Galaxy, as it displays a multi-ring-like structure when viewed in infrared radiation.

A lenticular galaxy is an intermediate form that has properties of both elliptical and spiral galaxies. These are categorized as Hubble type S0, and they possess ill-defined spiral arms with an elliptical halo of stars. (Barred lenticular galaxies receive Hubble classification SB0.)

NGC 5866, an example of a lenticular galaxy

In addition to the classifications mentioned above, there are a number of galaxies that can not be readily classified into an elliptical or spiral morphology. These are categorized as irregular galaxies. An Irr-I galaxy has some structure but does not align cleanly with the Hubble classification scheme. Irr-II galaxies do not possess any structure that resembles a Hubble classification, and may have been disrupted. Nearby examples of (dwarf) irregular galaxies include the Magellanic Clouds.

Dwarfs

Despite the prominence of large elliptical and spiral galaxies, most galaxies in the universe appear to be dwarf galaxies. These tiny galaxies are about one hundredth the size of the Milky Way, containing only a few billion stars. Ultra-compact dwarf galaxies have recently been discovered that are only 100 parsecs across.

Many dwarf galaxies may orbit a single larger galaxy; the Milky Way has at least a dozen such satellites, with an estimated 300–500 yet to be discovered. Dwarf galaxies may also be classified as elliptical, spiral, or irregular. Since small dwarf ellipticals bear little resemblance to large ellipticals, they are often called dwarf spheroidal galaxies instead.

Unusual Dynamics and Activities

Interacting

The average separation between galaxies within a cluster is a little over an order of magnitude larger than their diameter. Hence interactions between these galaxies are relatively frequent, and play an important role in their evolution. Near misses between galaxies result in warping distortions due to tidal interactions, and may cause some exchange of gas and dust.

The Antennae Galaxies are undergoing a collision that will result in their eventual merger

Collisions occur when two galaxies pass directly through each other and have sufficient relative momentum not to merge. The stars within these interacting galaxies will typically pass straight through without colliding. However, the gas and dust within the two forms will interact. This can trigger bursts of star formation as the interstellar medium becomes disrupted and compressed. A collision can severely distort the shape of one or both galaxies, forming bars, rings or tail-like structures.

At the extreme of interactions are galactic mergers. In this case the relative momentum of the two galaxies is insufficient to allow the galaxies to pass through each other. Instead, they gradually merge together to form a single, larger galaxy. Mergers can result in significant changes to morphology, as compared to the original galaxies. In the case where one of the galaxies is much more massive, however, the result is known as cannibalism. In this case the larger galaxy will remain relatively undisturbed by the merger, while the smaller galaxy is torn apart. The Milky Way galaxy is currently in the process of cannibalizing the Sagittarius Dwarf Elliptical Galaxy and the Canis Major Dwarf Galaxy.

Starburst

M82, the archetype starburst galaxy, has experienced a 10-fold increase in star formation rate as compared to a "normal" galaxy

Stars are created within galaxies from a reserve of cold gas that forms into giant molecular clouds. Some galaxies have been observed to form stars at an exceptional rate, known as a starburst. Should they continue to do so, however, they would consume their reserve of gas in a time frame lower than the lifespan of the galaxy. Hence starburst activity usually lasts for only about ten million years, a relatively brief period in the history of a galaxy. Starburst galaxies were more common during the early history of the universe, and, at present, still contribute an estimated 15 percent to the total star production rate.

Starburst galaxies are characterized by dusty concentrations of gas and the appearance of newly-formed stars, including massive stars that ionize the surrounding clouds to create H II regions. These massive stars also produce supernova explosions, resulting in expanding remnants that interact powerfully with the surrounding gas. These outbursts trigger a chain reaction of star building that spreads throughout the gaseous region. Only when the available gas is nearly consumed or dispersed does the starburst activity come to an end.

Starbursts are often associated with merging or interacting galaxies. The prototype example of such a starburst-forming interaction is M82, which experienced a close encounter with the larger M81. Irregular galaxies often exhibit spaced knots of starburst activity.

Active Nucleus

A portion of the galaxies we can observe are classified as active. That is, a significant portion of the total energy output from the galaxy is emitted by a source other than the stars, dust and interstellar medium.

The standard model for an active galactic nucleus is based upon an accretion disc that forms around a supermassive black hole (SMBH) at the core region. The radiation from an active galactic nucleus results from the gravitational energy of matter as it falls toward the black hole from the disc. In about 10 percent of these objects, a diametrically opposed pair of energetic jets ejects particles from the core at velocities close to the speed of light.

A jet of particles is being emitted from the core of the elliptical radio galaxy M87

Active galaxies that emit high-energy radiation in the form of x-rays are classified as Seyfert galaxies or quasars, depending on the luminosity. Blazars are believed to be an active galaxy with a relativistic jet that is pointed in the direction of the Earth. A radio galaxy emits radio frequencies from relativistic jets. A unified model of these types of active galaxies explains their differences based on the viewing angle of the observer.

Larger Scale Structures

Deep sky surveys show that galaxies are often found in relatively close association with other galaxies. Solitary galaxies that have not significantly interacted with another galaxy of comparable mass during the past billion years are relatively scarce. Only about 5 percent of the galaxies surveyed have been found to be truly isolated; however, these isolated formations may have interacted and even merged with other galaxies in the past, and may still be orbited by smaller, satellite galaxies. Isolated galaxies can produce stars at a higher rate than normal, as their gas is not being stripped by other, nearby galaxies.

On the largest scale, the universe is continually expanding, resulting in an average increase in the separation between individual galaxies (see Hubble's law). Associations of galaxies can overcome this expansion on a local scale through their mutual gravitational attraction. These associations formed early in the universe, as clumps of dark matter pulled their respective galaxies together. Nearby groups later merged to form larger-scale clusters. This on-going merger process (as well as an influx of infalling gas) heats the inter-galactic gas within a cluster to very high temperatures, reaching 30–100 million K. About 70–80 percent of the mass in a cluster is in the form of dark matter, with 10–30 percent consisting of this heated gas and the remaining few percent of the matter in the form of galaxies.

Seyfert's Sextet is an example of a compact galaxy group

Most galaxies in the universe are gravitationally bound to a number of other galaxies. These form a fractal-like hierarchy of clustered structures, with the smallest such associations being termed groups. A group of galaxies is the most common type of galactic cluster, and these formations contain a majority of the galaxies (as well as most of the baryonic mass) in the universe. To remain gravitationally bound to such a group, each member galaxy must have a sufficiently low velocity to prevent it from escaping. If there is insufficient kinetic energy, however, the group may evolve into a smaller number of galaxies through mergers.

Larger structures containing many thousands of galaxies packed into an area a few megaparsecs across are called clusters. Clusters of galaxies are often dominated by a single giant elliptical galaxy, known as the brightest cluster galaxy, which, over time, tidally destroys its satellite galaxies and adds their mass to its own.

Superclusters contain tens of thousands of galaxies, which are found in clusters, groups and sometimes individually. At the supercluster scale, galaxies are arranged into sheets and filaments surrounding vast empty voids. Above this scale, the universe appears to be isotropic and homogeneous.

The Milky Way galaxy is a member of an association named the Local Group, a relatively small group of galaxies that has a diameter of approximately one megaparsec. The Milky Way and the Andromeda Galaxy are the two brightest galaxies within the group; many of the other member galaxies are dwarf companions of these two galaxies. The Local Group itself is a part of a cloud-like structure within the Virgo Supercluster, a large, extended structure of groups and clusters of galaxies centred around the Virgo Cluster.

Multi-wavelength Observation

Color composite image of Centaurus A, revealing the lobes and jets emanating from the active galaxy's central black hole

After galaxies external to the Milky Way were found to exist, initial observations were made mostly using visible light. The peak radiation of most stars lies here, so the observation of the stars that form galaxies has been a major component of optical astronomy. It is also a favorable portion of the spectrum for observing ionized H II regions, and for examining the distribution of dusty arms.

The dust present in the interstellar medium is opaque to visual light. It is more transparent to far-infrared, which can be used to observe the interior regions of giant molecular clouds and galactic cores in great detail. Infrared is also used to observe distant, red-shifted galaxies that were formed much earlier in the history of the universe. Water vapor and carbon dioxide absorb a number of useful portions of the infrared spectrum, so high-altitude or space-based telescopes are used for infrared astronomy.

The first non-visual study of galaxies, particularly active galaxies, was made using radio frequencies. The atmosphere is nearly transparent to radio between 5 MHz and 30 GHz. (The ionosphere blocks signals below this range.) Large radio interferometers have been used to map the active jets emitted from active nuclei. Radio telescopes can also be used to observe neutral hydrogen (*via* 21 centimetre radiation), including, potentially, the non-ionized matter in the early universe that later collapsed to form galaxies.

Ultraviolet and X-ray telescopes can observe highly energetic galactic phenomena. An ultraviolet flare was observed when a star in a distant galaxy was torn apart from the tidal forces of a black hole. The distribution of hot gas in galactic clusters can be mapped by X-rays. The existence of super-massive black holes at the cores of galaxies was confirmed through X-ray astronomy.

Galaxy Formation and Evolution

Formation and Evolution

The study of galactic formation and evolution attempts to answer questions regarding how galaxies formed and their evolutionary path over the history of the universe. Some theories in this field have now become widely accepted, but it is still an active area in astrophysics.

Formation

Current cosmological models of the early Universe are based on the Big Bang theory. About 300,000 years after this event, atoms of hydrogen and helium began to form, in an event called recombination. Nearly all the hydrogen was neutral (non-ionized) and readily absorbed light, and no stars had yet formed. As a result this period has been called the "Dark Ages." It was from density fluctuations (or anisotropic irregularities) in this primordial matter that larger structures began to appear. As a result, masses of baryonic matter started to condense within cold dark matter halos. These primordial structures would eventually become the galaxies we see today.

Evidence for the early appearance of galaxies was found in 2006, when it was discovered that the galaxy IOK-1 has an unusually high redshift of 6.96, corresponding to just 750 million years after the Big Bang and making it the most distant and primordial galaxy yet seen. While some scientists have claimed other objects (such as Abell 1835 IR1916) have higher redshifts (and therefore are seen in an earlier stage of the Universe's evolution), IOK-1's age and composition have been more reliably established. The existence of such early protogalaxies suggests that they must have grown in the so-called "Dark Ages".

The detailed process by which such early galaxy formation occurred is a major open question in astronomy. Theories could be divided into two categories: top-down and bottom-up. In top-down theories (such as the Eggen–Lynden-Bell–Sandage [ELS] model), protogalaxies form in a large-scale simultaneous collapse lasting about one hundred million years. In bottom-up theories (such as the Searle-Zinn [SZ] model), small structures such as globular clusters form first, and then a number of such bodies accrete to form a larger galaxy. Modern theories must be modified to account for the probable presence of large dark matter halos.

Once protogalaxies began to form and contract, the first halo stars (called Population III stars) appeared within them. These were composed almost entirely of hydrogen and helium, and may have been massive. If so, these huge stars would have quickly consumed their supply of fuel and became supernovae, releasing heavy elements into the interstellar medium. This first generation of stars re-ionized the surrounding neutral hydrogen, creating expanding bubbles of space through which light could readily travel.

Evolution

I Zwicky 18 (lower left) resembles a newly-formed galaxy

Within a billion years of a galaxy's formation, key structures begin to appear. Globular clusters, the central supermassive black hole, and a galactic bulge of metal-poor Population II stars form. The creation of a supermassive black hole appears to play a key role in actively regulating the growth of galaxies by limiting the total amount of additional matter added. During this early epoch, galaxies undergo a major burst of star formation.

During the following two billion years, the accumulated matter settles into a galactic disc. A galaxy will continue to absorb infalling material from high velocity clouds and dwarf galaxies throughout its life. This matter is mostly hydrogen and helium. The cycle of stellar birth and death slowly increases the abundance of heavy elements, eventually allowing the formation of planets.

The evolution of galaxies can be significantly affected by interactions and collisions. Mergers of galaxies were common during the early epoch, and the majority of galaxies were peculiar in morphology. Given the distances between the stars, the great majority of stellar systems in colliding galaxies will be unaffected. However, gravitational stripping of the interstellar gas and dust that makes up the spiral arms produces a long train of stars known as tidal tails. An examples can be seen in NGC 4676.

The Milky Way galaxy and the nearby Andromeda Galaxy are moving toward each other at about 130 km/s, and depending upon the lateral movements the two may collide in about five to six billion years. Although the Milky Way has never collided with a galaxy as large as Andromeda before, evidence of past collisions of the Milky Way with smaller dwarf galaxies is increasing.

Such large-scale interactions are rare. As time passes, mergers of two systems of equal size become less common. Most bright galaxies have remained fundamentally unchanged for the last few billion years, and the net rate of star formation probably also peaked approximately ten billion years ago.

Future Trends

At present, most star formation occurs in smaller galaxies where cool gas is not so depleted. Spiral galaxies, like the Milky Way, only produce new generations of stars as long as they have dense molecular clouds of interstellar hydrogen in their spiral arms. Elliptical galaxies are already largely devoid of this gas, and so form no new stars.

The supply of star-forming material is finite; once stars have converted the available supply of hydrogen into heavier elements, new star formation will come to an end. The current era of star formation is expected to continue for up to one hundred billion years, and then the "stellar age" will wind down after about ten trillion to one hundred trillion years (10^{13}–10^{14} years), as the smallest, longest-lived stars in our astrosphere, tiny red dwarfs, begin to fade. At the end of the stellar age, galaxies will be composed of compact objects: brown dwarfs, white dwarfs that are cooling or cold ("black dwarfs"), neutron stars, and black holes. Eventually, as a result of gravitational relaxation, all stars will either fall into central supermassive black holes or be flung into intergalactic space as a result of collisions.

Galaxy Morphological Classification

Astronomers use the term 'morphology' to refer to the structural properties of galaxies. A galaxy's Hubble classification provides one way of describing its morphology, however, this classification scheme only considers the most prominent features: disks, bulges and bars. A more complete morphological classification of galaxies would include features such as extended stellar halos, warps, shells and tidal tails. Galaxies which clearly exhibit these morphological features are said to possess morphological peculiarities, and their Hubble type is usually followed by the suffix 'pec' (short for peculiar).

M87 has an elliptical morphology

NGC4526 has a lenticular morphology

M100 has a spiral morphology

NGC1313 has a hard to classify, irregular morphology

Both the prominent features of galaxies and morphological peculiarities can be explained by models of galaxy formation and evolution.

Hubble Classification

The Hubble classification of galaxies, also referred to as the 'tuning fork' diagram because of its shape, classes galaxies along three main lines into:

- Elliptical galaxies
- Spiral galaxies
- Barred Spiral Galaxies

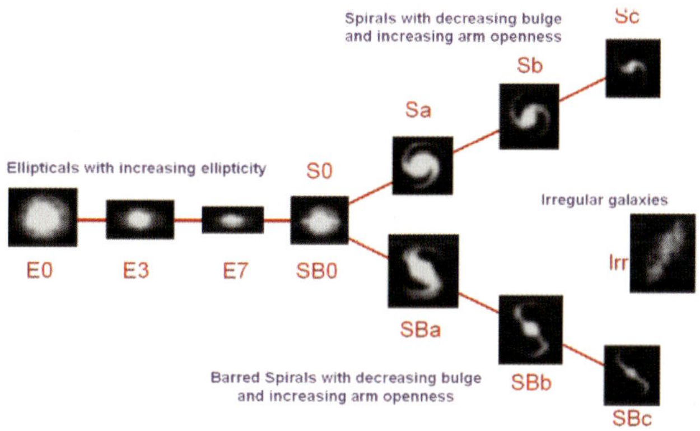

The Hubble Classification scheme for galaxies, often referred to as the "tuning fork" diagram

Edwin Hubble originally identified an evolutionary sequence for the galaxies (from early-type to late-type) as one moved from left to right across the diagram. Although this is now known to be a false interpretation, the terms 'early-type' and 'late-type' are still used regularly by astronomers in the manner described below, and when discussing broad galaxy types.

Elliptical Galaxies

Hubble's elliptical galaxies were classed according to the ellipticity of the galaxy, and given a Hubble type:

$$E = 10 \text{ x } (1 - b/a)$$

where a = semi-major axis and b = semi-minor axis of the ellipse. Observed values range from E0 (circular cross section – a spherical galaxy) to E7 (the most flattened). E0 are considered 'early-type' ellipticals and E7 are 'late-type' ellipticals.

S0/SB0 Galaxies

Located in the fork of the Hubble classification diagram and intermediate between the elliptical and spiral galaxies are the S0/SB0 galaxies. These galaxies show prominent bulges, but no spiral arms.

Spiral Galaxies

Spiral galaxies are classed from 'early-type' to 'late-type' according to the ratio of the luminosity of the bulge compared with the disk, and the amount of winding of the spiral arms. Type Sa (early) spiral galaxies have prominent bulges (bulge to disk luminosity ~0.3), tightly wound arms (pitch angle ~ 0.6), and the stars in the spiral arms are distributed very smoothly. Type Sc (late) spirals have the least prominent bulges (bulge-to-disk luminosity ~ 0.05), and loosely wound arms (pitch angle ~ 0.18) that are resolved into clumps of stars and HII regions.

Barred Spiral Galaxies

In bars, the spiral arms do not start directly from the bulge, but from an extended bar of stars that passes through the bulge. They share the same range of classifications as non-barred spirals – from Type SBa (early) to SBc (late) – according to the prominence of the bulge and the winding of the spiral arms.

Irregular Galaxies

A later extension to the Hubble classification was the inclusion of irregular galaxies in two classes: Irr I included irregular galaxies that showed some hint of organised structure (such as the Large and Small Magellanic Clouds), while Irr II were those irregulars that were completely disorganised.

Comparison of Hubble Types

Class	Absolute magnitude **(B-band)**	Mass **(**solar masses**)**	**Diameter (**kiloparsec**)**
Ellipticals	-8 to -23	10^7 to 10^{12}	~0.3 to 100s
Spirals	-16 to -23	10^9 to 10^{12}	5 to 100
Irregulars	-13 to -20	10^8 to 10^{10}	1 to 10

De Vaucouleurs System

The de Vaucouleurs system for classifying galaxies is a widely used extension to the Hubble sequence, first described by Gérard de Vaucouleurs in 1959. De Vaucouleurs argued that Hubble's two-dimensional classification of spiral galaxies based on the tightness of the spiral arms and the presence or absence of a bar did not adequately describe the full range of observed galaxy morphologies. In particular, he argued that rings and lenses are important structural components of spiral galaxies.

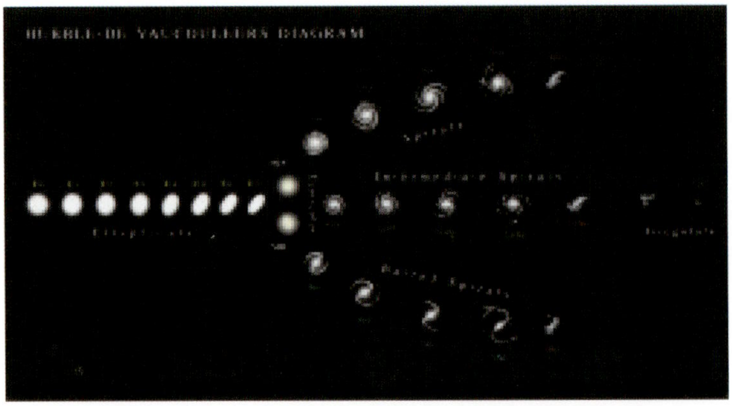

Hubble - de Vaucouleurs Galaxy Morphology Diagram

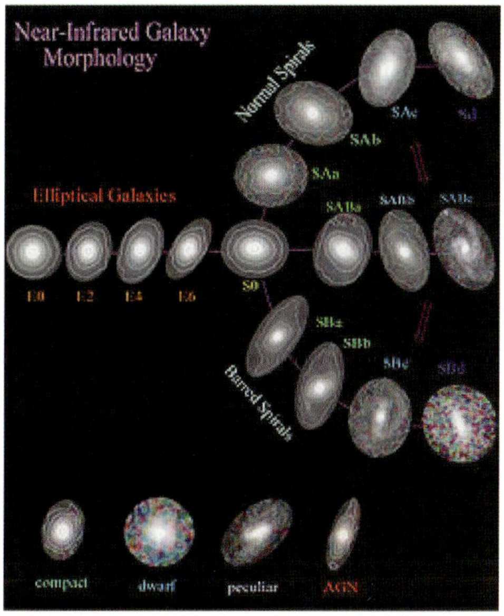

NGC 6782: a spiral galaxy (type SB(r)o/a) with three rings of different radii, as well as a bar

The de Vaucouleurs system retains Hubble's basic division of galaxies into ellipticals, lenticulars, spirals and irregulars. To complement Hubble's scheme, de Vaucouleurs introduced a more elaborate classification system for spiral galaxies, based on three morphological characteristics:

- Bars: Galaxies are divided on the basis of the presence or absence of a nuclear bar. De Vaucouleurs introduced the notation SA to denote spiral galaxies without bars, complementing Hubble's use of SB for barred spirals. He also allowed for an intermediate class, denoted SAB, containing weakly barred spirals. Lenticular galaxies are also classified as unbarred (SAo) or barred (SBo), with the notation So reserved for those galaxies for which it is impossible to tell if a bar is present or not (usually because they are edge-on to the line-of-sight).
- Rings: Galaxies are divided into those possessing ring-like structures (denoted '(r)') and those without rings (denoted '(s)'). So-called 'transition' galaxies are given the symbol (rs).
- Spiral arms: As in Hubble's original scheme, spiral galaxies are assigned to a class based primarily on the tightness of their spiral arms. The de Vaucouleurs scheme extends the arms of Hubble's tuning fork to include several additional spiral classes:
    - Sd (SBd) - diffuse, broken arms made up of individual stellar clusters and nebulae; very faint central bulge
    - Sm (SBm) - irregular in appearance; no bulge component
    - Im - highly irregular galaxy

  Most galaxies in these three classes were classified as Irr I in Hubble's original scheme. In addition, the Sd class contains some galaxies from Hubble's Sc class. Galaxies in the classes Sm and Im are termed the "Magellanic" spirals and irregulars, respectively, after the Magellanic Clouds. The Large Magellanic Cloud is of type SBm, while the Small Magellanic Cloud is an irregular (Im).

The different elements of the classification scheme are combined in the order in which they are listed to give the complete classification of a galaxy. For example, a weakly barred spiral galaxy with loosely wound arms and a ring is denoted SAB(r)c.

NGC 7793: a spiral galaxy of type SA(s)d

The Large Magellanic Cloud: a type SBm galaxy

Visually, the de Vaucouleurs system can be represented as a three-dimensional version of Hubble's tuning fork, with stage (spiralness) on the *x*-axis, family (barredness) on the *y*-axis, and variety (ringedness) on the *z*-axis.

Milky Way

Our Sun (a star) and all the planets around it are part of a galaxy known as the Milky Way Galaxy. A galaxy is a large group of stars, gas, and dust bound together by gravity. They come in a variety of shapes and sizes. The Milky Way is a large barred spiral galaxy. All the stars we see in the night sky are in our own Milky Way Galaxy. Our galaxy is called the Milky Way because it appears as a milky band of light in the sky when you see it in a really dark area.

It is very difficult to count the number of stars in the Milky Way from our position inside the galaxy. Our best estimates tell us that the Milky Way is made up of approximately 100 billion stars. These stars form a large disk whose diameter is about 100,000 light years. Our Solar System is about 25,000 light years away from the centre of our galaxy – we live in the suburbs of our galaxy.

Just as the Earth goes around the Sun, the Sun goes around the centre of the Milky Way. It takes 250 million years for our Sun and the solar system to go all the way around the centre of the Milky Way.

We can only take pictures of the Milky Way from inside the galaxy, which means we don't have an image of the Milky Way as a whole. Why do we think it is a barred spiral galaxy, then? There are several clues.

The first clue to the shape of the Milky Way comes from the bright band of stars that stretches across the sky (and, as mentioned above, is how the Milky Way got its name). This band of stars can be seen with the naked eye in places with dark night skies. That band comes from seeing the disk of stars that forms the Milky Way from inside the disk, and tells us that our galaxy is basically flat.

The band of the Milky Way galaxy can be seen at night in areas with dark skies. Here it is seen with several Atacama Large Millimeter/submillimeter Arra (ALMA) antenna.

An all-sky image shows the flat plane of the Milky Way galaxy

Several different telescopes, both on the ground and in space, have taken images of the disk of the Milky Way by taking a series of pictures in different directions – a bit like taking a panoramic picture with your camera or phone. The concentration of stars in a band adds to the evidence that the Milky Way is a spiral galaxy. If we lived in an elliptical galaxy, we would see the stars of our galaxy spread out all around the sky, not in a single band.

Another clue comes when astronomers map young, bright stars and clouds of ionized hydrogen in the Milky Way's disk. These clouds, called HII regions, are ionized by young, hot stars and are basically free protons and electrons. These are both important marker of spiral arms in other spiral galaxies we see, so mapping them in our own galaxy can give a clue about the spiral nature of the Milky Way. There are bright enough that we can see them through the disk of our galaxy, except where the region at the centre of our galaxy gets in the way.

There has been some debate over the years as to whether the Milky Way has two spiral arms or four. The latest data shows that it has four arms, as shown in the artist's illustration below.

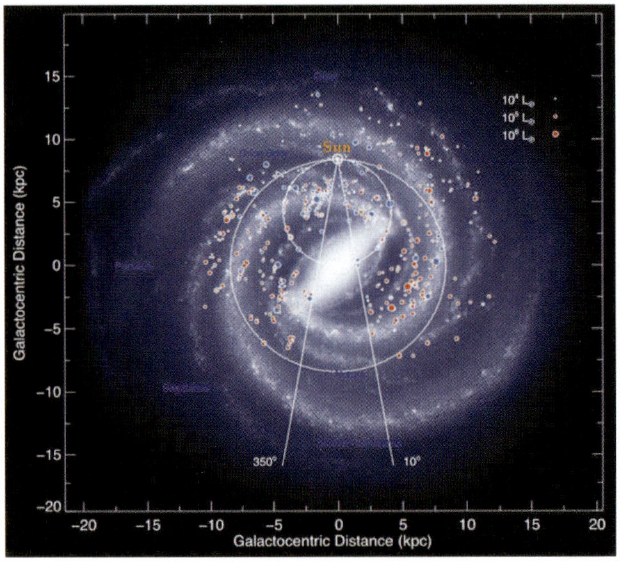

massive stars and HII regions" />
Since we can't get outside the Milky Way, we have to rely on markers of spiral arms like young, massive stars and ionized clouds. This artist's conception of the Milky Way's spiral structure is based on the measured distances of young, hot stars (shown in red) and ionized clouds of hydrogen gas (shown in blue)

Additional clues to the spiral nature of the Milky Way come from a variety of other properties. Astronomers measure the amount of dust in the Milky Way and the dominant colors of the light we see, and they match those we find in other typical spiral galaxies. All of this adds up to give us a picture of the Milky Way, even though we can't get outside to see the whole thing.

There are billions of other galaxies in the Universe. Only three galaxies outside our own Milky Way Galaxy can be seen without a telescope, and appear as fuzzy patches in the sky with the naked eye. The closest galaxies that we can see without a telescope are the Large and Small Magellanic Clouds. These satellite galaxies of the Milky Way can be seen from the southern hemisphere. Even they are about 160,000 light years from us. The Andromeda Galaxy is a larger galaxy that can be seen from the northern hemisphere (with good eyesight and a very dark sky). It is about 2.5 million light years away from us, but its getting closer, and researchers predict that in about 4 billion years it will collide with the Milky Way. , i.e., it takes light 2.5 million years to reach us from one of our "nearby" galaxies. The other galaxies are even further away from us and can only be seen through telescopes.

Our nearest large neighbor galaxy is the Andromeda galaxy

Galactic Centre

The Galactic Centre is the rotational centre of the Milky Way in the direction of the constellations Sagittarius, Ophiuchus and Scorpius. Since Earth is located within our galaxy's disk, our view of the Galactic Centre is blocked by layers of interstellar dust, making it difficult to study the centre through visible or ultraviolet wavelengths. Instead, we study the centre of the galaxy by observing at gamma ray, infrared and radio wavelengths and also by looking at the other galaxies' centre.

Our view of the Galactic Centre may be blocked, however we can still look for part of the centre of the Milky Way that includes the central bulge. The central area of the Milky Way can be seen during the late summer, preferably from an area that is far away from light pollution. By looking at the southern horizon, towards the constellation Sagittarius Arm, you can get a sense of the geometry of the Galactic Centre. The Sagittarius Arm, which spirals out of the the centre, actually blocks our direct view of the centre.

The exact distance from the Sun to the Galactic Centre is uncertain. The latest estimate yields a distance of 27,000 light.-years. An astronomical radio source called Sagittarius A* appears to be right at the centre of the Galactic Centre It is believed that Sagittarius A* is the location of the supermassive black hole, which is now the generally accepted view as what is at the centre of most galaxies.

Surrounding Sagittarius A* are thousands of stars. Compared to the population of stars around the Sun, the Galactic Centre is like a bustling city while the Sun is in a quieter neighbourhood. The Galactic Centre consists of many old main sequence stars and more than a hundred massive stars. These older generations of stars, formed more than 13 billion years ago, are classified as Population II. The younger generations of stars near the outskirt of the galaxy are classified as Population I.

Although many stars at the centre of the galaxy are believed to be old, some of these stars are the younger generation of stars. These younger stars entreat astronomers since they expect that the tidal force of the supermassive black hole would prevent star formations. This led astronomers to believer that stars are formed at the outskirt of the galaxy before migrating to the centre. Another scenario involves stars forming within the massive, compact gas of the accretion disk around the supermassive black hole.

Interacting Galaxy

An 'interacting galaxy' is one that is in the process of being affected by another galaxy. This is not an uncommon occurrence as galaxies are rarely found in isolation. Most are surrounded by a swarm of satellite galaxies and are themselves embedded in larger aggregates called groups or clusters.

Interactions can disturb, or even radically change, the morphologies of the galaxies involved, often inducing new bursts of star formation. The most common interaction is known as a 'fly-by' and involves two or more galaxies (which do not actually come into contact) approaching close enough that the gravitational field of each galaxy influences the gravitational fields of the others. These high speed encounters can create new structures such as warps or bars, or even tidal tails that extend well beyond the main body of the galaxy.

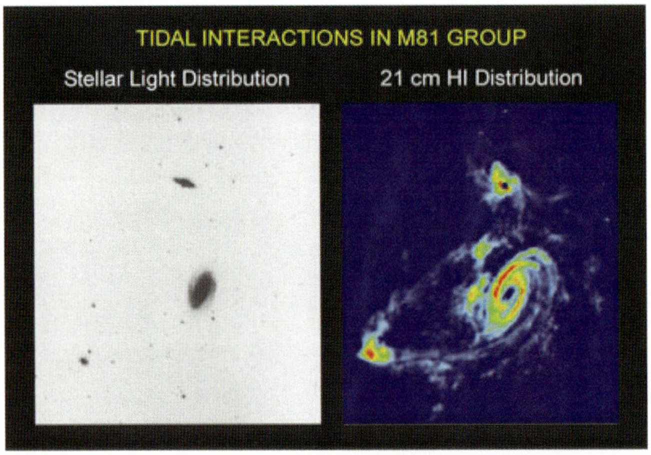

Optical and radio images of the interacting M81 group. Tidal tails of gas are clearly evident in the radio image

A spectacular example of a multiple interaction is the Messier 81 group of galaxies. The left-hand panel shows the optical view of M81 (the large dominant galaxy) with neighbouring galaxies NGC 3034 (above) and NGC 3077 (lower left). The right-hand panel shows the HI (radio) image which traces the distribution of cold gas in the system. This reveals long tidal tails of gas, not seen in the optical image, which connect the galaxies – a clear indication that all three galaxies are interacting with each other.

Another impressive example of an interacting galaxy is the Cartwheel galaxy. This galaxy was similar to our own Milky Way until quite recently, when a smaller galaxy (probably one of the galaxies on the right) ploughed through its centre. The interaction has created spokes of stars and gas, and a bright blue ring of new star formation. This has resulted in a wheel-like appearance for the galaxy and its descriptive name.

The Cartwheel galaxy has recently suffered an interaction, probably when one of the two galaxies to the right ploughed through its centre

Another form of interaction that can radically alter the pre-existing morphology of a galaxy is a merger event. A well-known example of an on-going major merger (i.e. one that involves galaxies with approximately equal masses) is the Antennae. This system exhibits two massive tidal tails thrown off by the gravitational interaction between the galaxies (left panel), and many new star clusters (right panel; bright blue knots) created through the interaction. Astronomers use the Toomre Sequence to visualise and track the progression of morphological changes that take place during major mergers.

The interacting galaxies NGC4038/39 – the Antennae

Of course, not all mergers involve galaxies of similar masses, and minor mergers (where one galaxy is substantially more massive than the other) are far more common. This image shows an on-going minor merger observed by Swinburne astronomers Duncan Forbes and Mike Beasley with the ACS instrument on the HST. It appears that the smaller satellite dwarf galaxy is being shredded by the larger galaxy, forming a large tidal tail which connects and extends beyond the two galaxies. Ultimately, the smaller galaxy will be completely disrupted and will merge with the larger galaxy.

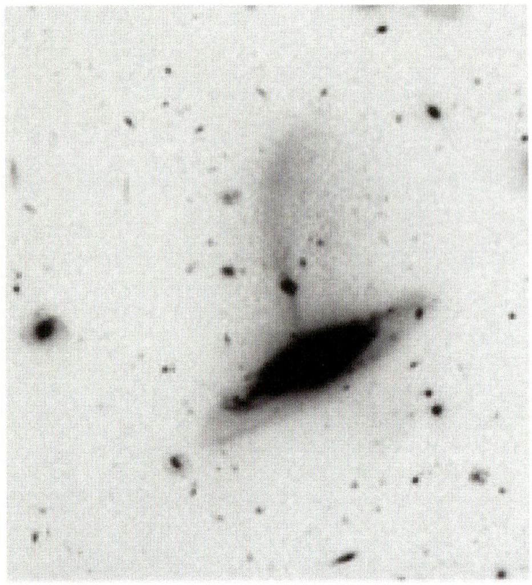

A satellite galaxy is 'shredded' by its parent galaxy

The Milky Way is currently undergoing minor mergers with both the Sagittarius and Canis Major dwarf galaxies. These galaxies are currently in the process of being shredded and absorbed by the Milky Way.

The large number of disturbed galaxies in these six regions of the Hubble Ultra Deep Field show that interactions between galaxies were even more common in the past than they are today

While it would seem that interactions are common in the local Universe, both models and observations of the distant Universe have shown that they were even more prevalent in the past. For example, six regions from the Hubble Ultra-Deep Field (shown right) contain an extraordinary number of interacting galaxies.

Consequently, over the 13 billion years since the Big Bang it seems highly probable that any individual galaxy will have suffered at least one major interaction with another galaxy, as well as several minor interactions. Interactions are therefore very important in the development of galaxies, though the degree to which galaxy formation and evolution is driven by interactions is still being investigated.

Magellanic Clouds

Magellanic Cloud, either of two satellite galaxies of the Milky Way Galaxy, the vast star system of which Earth is a minor component. These companion galaxies were named for the Portuguese navigator Ferdinand Magellan, whose crew discovered them during the first voyage around the world (1519–22). The Magellanic Clouds were recognized early in the 20th century as companion objects to the Milky Way Galaxy. When American astronomer Edwin Hubble established the extragalactic nature of what are now called galaxies, it became plain that the Magellanic Clouds had to be separate systems.

Globular cluster NGC 1850 in the Large Magellanic Cloud.Most of the cluster consists of yellow stars; the bright white stars are members of a second, open cluster about 200 light-years beyond NGC 1850. This picture is a composite of images taken by the Hubble Space Telescope

The Magellanic Clouds are irregular galaxies that share a gaseous envelope and lie about 22° apart in the sky near the south celestial pole. One of them, the Large Magellanic Cloud (LMC), is a luminous patch about 5° in diameter, and the other, the Small Magellanic Cloud (SMC), measures less than 2° across. The Magellanic Clouds are visible to the unaided eye in the Southern Hemisphere, but they cannot be observed from most northern latitudes. The LMC is about 160,000 light-years from Earth, and the SMC lies 190,000 light-years away. The LMC and SMC are 14,000 and 7,000 light-years in diameter, respectively smaller than the Milky Way Galaxy, which is about 140,000 light-years across.

Small Magellanic CloudInfant stars in the Small Magellanic Cloud

The Magellanic Clouds were formed at about the same time as the Milky Way Galaxy, approximately 13 billion years ago. They are presently captured in orbits around the Milky Way Galaxy and have experienced several tidal encounters with each other and with the Galaxy. They contain numerous young stars and star clusters, as well as some much older stars. One of these star clusters contains R136a1, the most massive star known, with a mass 265 times that of the Sun.

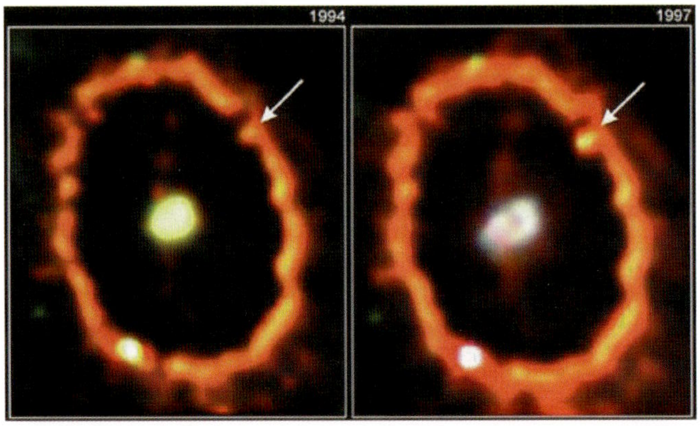

A knot in the central ring of Supernova 1987A, as observed by the Hubble Space Telescope in 1994 (left) and 1997 (right). The knot is caused by the collision of the supernova's blast wave with a slower-moving ring of matter it had ejected earlier. The bright spot on the lower left is an unrelated star

The Magellanic Clouds serve as excellent laboratories for the study of very active stellar formation and evolution. For example, the Tarantula Nebula (also called 30 Doradus) is an immense ionized-hydrogen region that contains many young, hot stars. The total mass of 30 Doradus is about one million solar masses, and its diameter is 550 light-years, making it the largest region of ionized gas in the entire Local Group of galaxies. With the Hubble Space Telescope it is possible for astronomers to study the kinds of stars, star clusters, and nebulae that previously could be observed in great detail only in the Milky Way Galaxy.

Characteristics

The Large Magellanic Cloud (LMC) and the neighboring the Small Magellanic Cloud (SMC) are starry regions that orbit our galaxy, and look conspicuously like detached pieces of the Milky Way. Though they are separated by 21 degrees in the night sky – about 42 times the width of the full moon – their true distance is about 75,000 light-years from each other.

Ultraviolet view of the Large Magellanic Cloud from Swift's Ultraviolet/Optical Telescope

The Large Magellanic Cloud is located about 160,000 light-years from the Milky Way, in the constellation Dorado. This makes it the 3rd closest galaxy to us, behind the Sagittarius Dwarf and Canis Major Dwarf galaxies. Meanwhile, the Small Magellanic Cloud is located in the constellation of Tucana, about 200,000 light-years away.

The LMC is roughly twice the diameter of the SMC, measuring some 14,000 light-years across vs. 7,000 light years (compared to 100,000 light years for the Milky Way). This makes it the 4th largest galaxy in our Local Group of galaxies, after the Milky Way, Andromeda and the Triangulum Galaxy. The LMC is about 10 billion times as massive as our Sun (about a tenth the mass of the Milky Way), while the SMC is equivalent to about 7 billion Solar Masses.

In terms of structure, astronomers have classified the LMC as an irregular type galaxy, but it does have a very prominent bar in its centre. Ergo, it's possible that it was a barred spiral before its gravitational interactions with the Milky Way. The SMC also contains a central bar structure and it is speculated that it too was once a barred spiral galaxy that was disrupted by the Milky Way to become somewhat irregular.

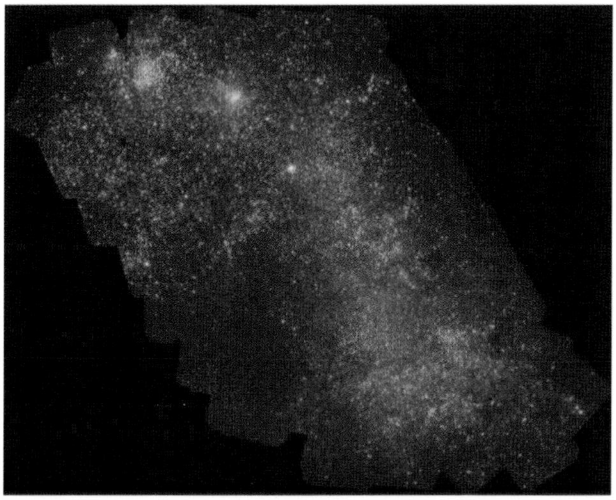

The Small Magellanic Cloud as seen by Swift's Ultraviolet/Optical Telescope. This composite of 656 separate pictures has a cumulative exposure time of 1.8 days

Aside from their different structure and lower mass, they differ from our galaxy in two major ways. First, they are gas-rich – meaning that a higher fraction of their mass is hydrogen and helium – and they have poor metallicity, (meaning their stars are less metal-rich than the Milky Way's). Both possess nebulae and young stellar populations, but are made up of stars that range from very young to the very old.

In fact, this abundance in gas is what ensures that the Magellanic Clouds are able to create new stars, with some being only a few hundred million years in age. This is especially true of the LMC, which produces new stars in large quantities. A good example of this is it's glowing-red Tarantula Nebula, a gigantic star-forming region that lies 160,000 light-years from Earth.

References

- Binney, J.; Merrifield, M. (1998). Galactic Astronomy. Princeton: Princeton University Press. ISBN 978-0-691-02565-0
- Galaxy, Formation and evolution: newworldencyclopedia.org, Retrieved 28 May 2018
- De Vaucouleurs, Gérard (April 1963). "Revised Classification of 1500 Bright Galaxies". Astrophysical Journal Supplement. 8: 31. Bibcode:1963ApJS....8...31D. doi:10.1086/190084
- Magellanic-Cloud: britannica.com, Retrieved 11 May 2018
- What-are-magellanic-clouds-30537: universetoday.com, Retrieved 31 March 2018

Permissions

All chapters in this book are published with permission under the Creative Commons Attribution Share Alike License or equivalent. Every chapter published in this book has been scrutinized by our experts. Their significance has been extensively debated. The topics covered herein carry significant information for a comprehensive understanding. They may even be implemented as practical applications or may be referred to as a beginning point for further studies.

We would like to thank the editorial team for lending their expertise to make the book truly unique. They have played a crucial role in the development of this book. Without their invaluable contributions this book wouldn't have been possible. They have made vital efforts to compile up to date information on the varied aspects of this subject to make this book a valuable addition to the collection of many professionals and students.

This book was conceptualized with the vision of imparting up-to-date and integrated information in this field. To ensure the same, a matchless editorial board was set up. Every individual on the board went through rigorous rounds of assessment to prove their worth. After which they invested a large part of their time researching and compiling the most relevant data for our readers.

The editorial board has been involved in producing this book since its inception. They have spent rigorous hours researching and exploring the diverse topics which have resulted in the successful publishing of this book. They have passed on their knowledge of decades through this book. To expedite this challenging task, the publisher supported the team at every step. A small team of assistant editors was also appointed to further simplify the editing procedure and attain best results for the readers.

Apart from the editorial board, the designing team has also invested a significant amount of their time in understanding the subject and creating the most relevant covers. They scrutinized every image to scout for the most suitable representation of the subject and create an appropriate cover for the book.

The publishing team has been an ardent support to the editorial, designing and production team. Their endless efforts to recruit the best for this project, has resulted in the accomplishment of this book. They are a veteran in the field of academics and their pool of knowledge is as vast as their experience in printing. Their expertise and guidance has proved useful at every step. Their uncompromising quality standards have made this book an exceptional effort. Their encouragement from time to time has been an inspiration for everyone.

The publisher and the editorial board hope that this book will prove to be a valuable piece of knowledge for students, practitioners and scholars across the globe.

Index

A

Asteroid Belt, 106, 108-109, 112-115, 133-135, 161, 166-168, 172, 177, 181, 206

B

Barred Spiral Galaxy, 120, 216, 228-229, 237

Big Bang, 3-6, 19, 87, 222, 235

C

Canis Major, 218, 234, 237

Ceres, 108-110, 113, 121, 155, 165-166, 168, 170, 172, 174, 176-177, 186, 206-208

Chemical Evolution Models, 18-19, 22

Chromosphere, 83, 85, 89-92, 105

Circular Orbit, 25, 27, 29, 31, 37-38, 42, 48, 63-64, 66-67, 69-70

Comet, 39, 62-63, 108, 115, 117, 121, 132, 134, 136, 141, 144, 164, 166, 178, 185, 203

Convection Zone, 21, 85, 88, 97-105

Corona, 83, 85, 89-90, 92

D

De Vaucouleurs System, 226-228

Dwarf Galaxy, 218, 234

Dwarf Planet, 108-109, 113, 116, 119, 165-166, 168, 172, 204-208

E

Elliptical Galaxy, 214-215, 218, 221, 230

Elliptical Orbit, 25, 27, 35, 40, 42, 48, 67, 119, 166

Escape Velocity, 28-30, 35, 62, 68, 120

Exoplanet, 151, 163, 179, 182, 186, 194-196, 198, 200-201

Extrasolar Planets, 120, 125-126, 151-153, 173, 178-179, 182, 185-186, 195-196, 198

G

Galactic Centre, 107, 120, 231-232

Gas Giants, 114-115, 120, 144, 155-156, 158, 160-161, 170, 175-176, 178-179, 185-188, 193, 196, 201-202

Giant Planet, 134, 140, 153, 159, 187, 191, 194, 196

Gravitational Microlensing, 159, 180, 199

Gravity Assist, 24, 38, 59-61, 63, 69-70

H

Habitable Planet, 204

Habitable Zones, 155-156, 187, 197

Heliopause, 106-107, 119, 121

Hubble Classification, 214-217, 224-226

Hyperbolic Orbit, 33, 35, 54

I

Ice Giants, 114, 170, 176, 188, 193

Inner Planets, 106, 111-112, 114, 117, 141

Irregular Galaxies, 217, 219, 226, 235

K

Kepler's Equation, 34-35, 47-50

Kuiper Belt, 106, 109-110, 115-119, 121-122, 127-128, 134, 141, 161, 164-168, 172, 177, 181, 204, 206, 208

L

Law of Gravitation, 39, 42, 46

Laws of Planetary Motion, 24, 27-28, 39, 109

M

Magellanic Cloud, 212, 227-228, 235-237

Magnetic Field, 2, 85-86, 90-91, 93, 95-99, 101-103, 105-107, 111-112, 122, 129, 183-184, 189, 192-193, 201-202

Magnetism, 85, 91

Magnetosphere, 107, 183-184, 192, 202

Meridional Flow, 85, 98, 101, 103-104

Milky Way, 62, 83, 107, 119-120, 124, 157, 159, 161, 168, 180, 196-197, 210-214, 217-218, 221, 223-224, 228-231, 233-238

Morphology, 169, 214, 217-218, 223-224, 226, 233

N

Nebulae, 1, 3, 5, 125, 212-213, 227, 236, 238

Neutrino, 5-6, 11, 13, 15, 18, 20, 22

Neutron Capture Process, 12, 14

Nice Model, 167-168, 194

Nuclear Astrophysics, 1, 11, 14, 19

Nucleosynthesis, 13, 16-19, 22

O

Oort Cloud, 106, 108, 115, 117-118, 161, 164-168

Orbital Elements, 24, 26, 29, 36, 63

Orbital Period, 27, 31, 42-43, 71, 119, 159

Orbital Velocity, 24, 29-30, 33-34, 51-54

P

Pair Production, 77-79

Patched Conic Approximation, 35, 54

Photodisintegration, 16, 77, 79

Photometry, 2, 152, 199

Photosphere, 20, 83-85, 87-88, 90-92, 105

Planetary-mass Object, 181, 204

Plasma, 4, 83, 85-87, 89-90, 92, 97-98, 107, 119-120, 122-123, 129, 164, 210

Polarimetry, 2, 199

R

Radial Velocity, 199

Radiation, 1-3, 5, 7-9, 11, 17, 74, 76, 79-80, 84, 86-88, 94, 107-108, 122, 128-130, 137, 142-143, 153, 178, 183, 192, 194, 203, 213, 216, 219-222

Ring System, 106, 114-115, 144-145, 202

Rogue Planet, 185, 203

S

Scattered Disc, 106, 109-110, 117-119, 161, 177, 206

Solar Core, 83, 86-87, 100, 102, 105

Solar Cycle, 83, 90, 93-94, 96, 105

Solar Dynamo, 83, 94, 96-98, 100, 102-103, 105

Solar System Bodies, 109, 112-113, 115-116, 131, 135, 174

Spectroscopy, 2-3, 6, 16-17, 152, 158

Spiral Galaxy, 120, 213-216, 227-230, 237

Starburst, 210, 214, 218-219

Stellar Structure, 7-9

Sub-brown Dwarf, 157, 181, 185

T

Terrestrial Planet, 112, 158, 178, 185, 187

The Oberth Effect, 38, 62, 68

Thermal Regime, 128, 131, 141

Transfer Orbit, 24, 35, 37, 55-56, 61, 63-70

V

Volcanism, 114, 144, 185, 202